The Silent Rape Epidemic

How the Finns Were Groomed to Love Their Abusers

Edward Dutton

ISBN: 9781799003649

Thomas Edward Press

The Silent Rape Epidemic: How the Finns Were Groomed to Love Their Abusers

In 2003, the northern Finnish city of Oulu was what foreigners imagined such a city would be: calm, quiet and very 'Finnish.' By December 2018, the whole country was convulsed by Oulu's Muslim Child Grooming Scandal. A gang of Arab 'refugees' were up in court for the sexual assault and rape of under age Oulu girls and it transpired that Finland's politicians and mainstream media had done everything possible – guided by the divine light of Multiculturalism – to cover it up. How could a country like Finland – stereotyped as nationalistic and culturally conservative – have embraced the Multicultural suicide cult so rapidly? In *The Silent Rape Epidemic,* Dutton argues that the explanation lies in Finns' evolutionary past. Adapted to extreme cold and predictability, they are fiercely intelligent, extraordinarily cooperative, highly rule following, silent, and socially anxious. This makes them naively trusting, conformist and prone to imitating others. Perfectly adapted to their environment, these traits are combined with a miniscule gene pool, meaning very few intelligence or personality outliers who might dare to challenge the Spirit of the Age. The intense Darwinian selection to which they've been subjected, and their late industrialization, has delayed the rise of nihilism, which arises when nations industrialize and selection pressures are weakened. But the nature of Finnishness means that Multiculturalism has now overwhelmed Finland with astonishing speed. However, all is not lost. Dutton finds that if the Finns truly want to save the *Suomi* of *sauna* and Sibelius, then they are, ironically, uniquely well-adapted to *reverse* their country's destruction.

Contents

About the Author

Edward Dutton is an independent scholar and writer based in Oulu in northern Finland. Dutton originally read Theology at Durham University, before completing a PhD in Religious Studies at Aberdeen University. In 2012, he made the move from Religious Studies and cultural anthropology to evolutionary psychology and has never looked back. Dutton has published in leading psychology journals including *Intelligence, Personality and Individual Differences,* and *Journal of Biosocial Science.* Dutton is editor-in-chief of the academic journal *Mankind Quarterly,* which aims not to shy away from contentious scientific areas. He is Adjunct Professor (Docent) of the Anthropology of Religion and Finnish Culture at Oulu University, academic consultant to a research group at King Saud University in Saudi Arabia, and has been a guest researcher at Umeå University in Sweden and Leiden University in the Netherlands. His other books include (with Michael Woodley of Menie) *At Our Wits' End: Why We're Becoming Less Intelligent and What It Means for the Future* (Imprint Academic, 2018); *How to Judge People by What They Look Like* (Thomas Edward Press, 2018); *J. Philippe Rushton: A Life History Perspective* (Thomas Edward Press, 2018) and *Meeting Jesus at University: Rites of Passage and Student Evangelicals* (Routledge, 2016). His research has been reported worldwide including in *Daily Telegraph, The Sun, Le Monde, Huffington Post* and *Newsweek.* Dutton's hobbies include genealogy, history and Indian cooking. Dutton vlogs on YouTube at 'The Jolly Heretic.'

Acknowledgements

I would like to thank Prof Guy Madison for kindly reading through and commenting on earlier versions of this book. I am indebted to Miss Tara McCarthy for designing the cover and for her ideas regarding the book's title. Herr David Becker and Gospodin Vladimir Shibaev have worked with me on my research into the Finns as have Dr Jan te Nijenhuis, Dr Eka Roivainen, Dr Michael Woodley of Menie, and Prof Dimitri van der Linden, so I would like to extend my thanks to them. I am, in addition, very grateful to Senor Davide Piffer for sharing his research findings with me.

Parts of this book have previously appeared in or are in press in: Dutton, E. & Woodley of Menie, M.A (2018) *At Our Wits' End: Why We're Becoming Less Intelligent and What It Means for the Future.* Exeter: Imprint Academic; Dutton, E. (In Press). *Race Differences in Ethnocentrism.* Budapest: Arktos; Dutton, E. (2019). Can You Marry a Foreigner Yet Still Be Ethnocentric? A Response to Cofnas' Criticisms of Kevin MacDonald's *Culture of Critique. Mankind Quarterly;* Dutton, E. (In Press). *Churchill's Headmaster: The 'Sadist' Who Nearly Saved the British Empire.* Manticore Press; Dutton, E. (October 2009). Finland and the Cold War. *History Today,* and Dutton, E. (2008). Battling to be European: Myth and the Finnish Race Debate. *Antrocom: Online Journal of Anthropology*, 5: 1.

Edward Dutton
7th March 2019, Oulu

'As the sun set, we revelled in the glories visible from our balcony, and thoroughly enjoyed the charms of the Northern night. Our souls were steeped in that great silence.'

Mrs Alec Tweedie, Through Finland in Carts, 1898.

Chapter One

Northern Finland's Muslim Grooming Epidemic

1. The Silent Finns

In July 2003, I sat on a plane leaving Heathrow airport. I was on my way to Helsinki to visit my Finnish girlfriend, whom I would marry two years later. Just over a year earlier, I'd been to Norway. I'd occupied myself, on the 90-minute flight, by chatting to the friendly butcher who was seated next to me. His English, though slightly broken and Transatlantic-sounding, had been very good, especially as he was a butcher and thus, presumably, had absolutely no reason to ever speak any language other than Norwegian. I knew that the Finnish language was completely different from English. It was Finno-Ugric – a peculiar linguistic group which includes Hungarian, Estonian and the dialects of the Saami reindeer herders of Lapland – rather than Indo-European, like English and Norwegian. My language and that of the butcher's were so similar that they'd have been mutually understandable a thousand years ago (Freeborn, 1998, p.149). But, even so, Finland was a highly advanced country with a language that nobody else spoke. Other than their strange tongue, why would they be significantly different from the Norwegians in any way? I was so confident that they would not be that I hadn't bothered bringing my disc-man or even a book to read. I was certain that this flight was going to 'fly by' while I chatted to some friendly Finn.

But the atmosphere alone should have told me that this carefully laid plan was going to go awry. Seemingly everyone on the plane was Finnish except me. And I was very aware of how quiet it was. It was almost silent. Nobody was talking to anybody else and if they were it was little more than a whisper. I hadn't noticed anything like this on the flight to Oslo a year before. And the people's faces seemed to be expressionless; inscrutable – as if they were deep in some kind of trance. I leant over to my neighbour, a blonde-haired man of about 40 with glasses and a small nose that stuck up so that you could see the nostrils from the front. As I did so, I also realised that quite a lot of the passengers were wearing glasses and wearing them atop noses of precisely this (in England) very unusual kind.

'Do you speak English?' I enquired, in a deliberately jolly-sounding way.

'I speak only very little,' he replied: monotonous, nervous, staccato.

I'd found that in Norway pretty much anyone under the age of 50 could get by in English and if they were my age – at that time, 22 – they'd speak English, essentially, fluently. As I attempted to talk to more people, a process which continued once I arrived in Finland, it emerged that this was not the case in the land of the Moomins and Father Christmas. This seemed to be due to a combination of factors: extreme shyness of an almost Japanese kind, the language having no relationship with English, and simply that they weren't as advanced – or should that be 'enriched'? – in terms of Globalization as the Norwegians were.

I'd been in Finland over a week when, on the way to Lapland, I passed through the north west coastal city of Oulu; an industrial place which, on that hot day in July, stank like a blocked drain. I was later to discover that this was due to the city's paper factory. It was in Oulu (pronounced 'O-loo') that it really hit me that I had gone back in time. Oulu had been founded as a city in 1605 and, like all Finnish cities at the time, it was Swedish-speaking. Finland-Swedes, at that time, were about 17% of Finland's population (now down to 5.4%) and the overwhelming majority of its elite (Taagepera, 2000). By 2003, the only evidence there'd ever been a Finland-Swede presence in Oulu was in the cemetery but the city they'd established had grown enormously. It had a population of about 200,000. There were factories, industrial estates; technology parks. But, despite this, everybody there was Finnish. I didn't see a single non-white face. There were no Indian restaurants; merely a handful of pizzerias and a couple of Chinese. Popping into a local shop, I noticed that the liquorice wrappers were decorated with Golliwogs – practically outlawed in Politically Correct Britain in the 1970s – and there was also a kind of chocolate teacake called, in direct translation, a 'Nigger Kisser.' This was charmingly adorned with a male and female, both stereotyped African tribespeople, kissing. We went to meet Päivi, a school friend of my Finnish girlfriend. There was always the initial taciturnity and timorousness when I met these people but, eventually, they'd open up. Finns I spoke to, including Päivi, were very happy about this lack of 'diversity.' Sweden was

'Multicultural: it 'accepted all these immigrants.' Finns I spoke to adamantly didn't want that to happen to Finland, because they saw what it led to in Sweden: crime, conflict, lack of trust. Already, there were some non-white immigrants in Helsinki, it was causing serious social strife, and a former professional wrestler called Tony Halme (1963-2010) was campaigning against it, having become a member of parliament on the back of it (Kuisma, 2013). The Finnish 'far right' was miniscule, I was told, because all of Finland's political parties were, by the standards of 'Europe' - many Finns didn't include themselves in 'Europe' – 'racist.'

Also, everyone my girlfriend introduced me to was, by English standards at least, religious: baptized, confirmed; paid up members of the Finnish Lutheran Church as were about 85% of Finns at the time. Her friends were undergraduate students at far more advanced ages than people were in the UK; beneficiaries of there being no tuition fees and there being student grants. I also began to notice evidence of high levels of trust in this part of the country: a contraption by the side of the road which sold tomatoes and trusted that you'd leave the right money for the right amount; trains that you could get on to without purchasing a ticket, the conductor making his way through to flocks of Finns happily volunteering to buy them; young girls living in villages thumbing lifts to and from the city centre on a Saturday night. This must have been what England was like in the 1960s. It was trusting precisely because everybody was Finnish. It has been shown that immigration reduces trust even between the natives, because there are suddenly outsiders to whom treacherous natives can defect and, naturally, the immigrants, being different, will not be trusted (Putnam, 2007). Religious people tend to be more trustworthy. They are higher in the character traits Conscientiousness (rule-following) and Agreeableness (altruism and empathy) even in societies which are not especially religious (Gebauer et al., 2014). This is because being religious is associated with a very high level of empathy, meaning a deep interest in, and ability to read, external signals of the feelings of others. This is so pronounced that it extends to reading signs of feelings into the world itself, meaning that the world is understood to be the reflection of a mind: God's mind. This is why autistics, who are very low in empathy, tend not to believe in God (Norenzayan et al., 2012). This personality-religiousness relationship is, however, symbiotic, because once people feel that a moral god is watching them they will behave in a more prosocial way, making them more trustworthy (Norenzayan & Shariff, 2008). Even the free higher education with a grant, which my parents' generation had enjoyed in the UK, likely reflected the high level of trust as well as the low level of ethnic diversity which elevated that high level of trust. In 2003, as far as I could see, the Finns were, genetically, one big family, helping each other out. And it has been shown that the more ethnically diverse a society is, the less likely a welfare state is to be maintained (Putnam, 2007), because people are motivated by the desire to help their genetic kin: family, kinship group, ethnic group. They invest more resources in people the more genetically similar they are to themselves, because it is a way of indirectly passing on their genes (Salter, 2007). This model – that selection works on a number of levels – is known as 'Multi-Level Selection Theory.'

It occurred to me, looking at newspapers and asking my girlfriend what was being reported, just what a boring place Oulu – and Finland in general – was to live in. Very little crime, very little conflict; very little to report: Everyone eating the same kind of food, wearing the same kind of clothes, following the same cultural traditions. The key cultural tradition seemed to be sitting very close together, naked, in a hot room: they called it 'sauna.' But Oulu was trusting. Oulu was safe. Oulu was predictable. Life had certainty and, therefore, unless you thought about it too deeply, meaning, for this is what predictable rituals provide people with (Boyer, 2001). Oulu was the perfect place to raise a child, boy or girl. It was so undramatic and peaceful that I understood, for the first time, the brilliance of *Monty Python's* satirical song about the country: 'Finland, Finland, Finland/ The country where I want to be/ Pony trekking or camping / Or just watching TV. . .'

Just two years later, when I moved to Oulu, Muslims had come . . . and unthinkable things had happened. This wasn't, in fact, the first time that gangs of foreign marauders had come to Oulu and interfered with its young girls. In 1377, when Oulu was just a village, it was raided by Novgorod bandits who took 'a large number of prisoners.' This was a kidnapping mission and the 'prisoners' would have been sold into slavery, slavery still being legal in Eastern Europe at the time (Korpela, 2018, p.137). These slaves would have been bought and sold and it has been documented that between the Middle Ages and the sixteenth century small numbers of Finnish slaves ended up in the Muslim-run Khanate of Crimea every year.

Finnish female slaves (mostly abducted from Karelia in the east of the country), regarded as exotic, were in particularly high demand. Due to their blonde hair and pale skin, they fetched very high prices (Korpela, 2014). But this historical connection between Oulu and Islam was long forgotten by 2003. In 2003, Oulu was Finnish, almost entirely Finnish.

By 2018, however, Oulu girls weren't making their way to the Khanate as sex slaves. The Khanate, and its attendant sexual exploitation of infidel females, had settled in Oulu. Just 15 years after my first trip to the city, Oulu was little different from any other industrial town in Western Europe in terms of its problem with gangs of Muslim men and their horrifying activities. What they'd done in Oulu had even hit the international press. *Monty Python* had sung that Finland was 'So sadly neglected/And often ignored.' Sadly, this was no longer the case. How could anyone ignore Finland, let alone Oulu, now? It raised frightening, yet crucial, questions. How could Oulu have changed from 'Finland' to Oulu-stan? And how could Finland possibly have changed so quickly?

2. The List of Names

'Rahmani Gheibali (rape), Yosefi Shiraqa (aggravated sexual abuse of a child), Mirzad Javad (aggravated sexual abuse of a child), Barhum Abdullhadi (rape), Humad Osman Ahmed Mohamed (rape), Mohamed Ali Osman (aggravated sexual abuse of a child) . . .' So the list of names of men up in court to be tried on 28th November 2018 continued. Junes Lokka, a computer programmer and an independent Finnish nationalist councillor in Oulu stared at his screen. They'd been right. The man who had contacted him on 30th November to tell him that his daughter and step-daughter had been molested by Muslim men wasn't exaggerating. In fact, it was far worse than he'd thought. Lokka had uncovered the systematic grooming and rape of Oulu underage girls by the Muslim men who had come among them as refugees. And the authorities were ensuring that nobody knew anything about it.

The local newspaper in the so-called 'Capital of Northern Scandinavia,' *Kaleva,* had long ago fallen to the Social Justice Warriors. Despite it being named after the mythical eastern homeland of the Finns, celebrated in its national epic *Kalevala, Kaleva* was run by just the kind of people who disdained the

15

ordinary Finnish 'folk;' their pride in holding out against the Soviets in the Winter War of 1939, their simple desire for a picturesque wooden house close to a lake; their contentment living in a society where people mostly think in the same way, trust each other, and in which the clock has to be turned back 30 years or so upon arrival, except when it comes to technology.

For over a decade, this newspaper – in tacit collusion with the police and the political class – had been suppressing what was happening in this industrial city of paper mills, computer game companies and, for a long time, Nokia. In the summer of 2005, not long after Oulu council had begun accepting Muslim 'refugees,' a 30-year-old worker for the Finnish Lutheran Church had been naïve enough to get talking to a group of Muslim men in a bar and, worse still, go back to their flat. She was rewarded for her friendliness towards these guests by having her clitoris cut off with a pair of scissors which were then inserted into her vagina (*Turun Sanomat*, 1st November 2005). Thereafter, desperate that ordinary Finnish people should accept 'Multiculturalism,' the names of Muslim rapists were no longer reported by newspapers. Eventually, they were also redacted from police and court press releases as well. In 2006, a number of girls were gang raped by Muslims in parks in the centre of the city. More girls were gang-raped by bringers of diversity in Oulu parks in 2007.

In December 2007, a 26-year-old woman was walking home one night through the Otto Karhi Park in the centre of the town. There was nothing unusual about young Oulu females walking home, alone and perhaps the worse for drink, late at night. Apart from intermittent fires burning down all of the wooden buildings and a skirmish during the country's Civil War of 1918, Oulu had never been a particularly eventful place. But it was safe. On this night, however, there was a Middle Eastern 'refugee' in the park, resulting in the young woman being raped. When national tabloid *Ilta-Lehti* ran an article on Oulu's rapes in 2008, it conveniently omitted the fact that all of these sexual assaults had been carried out by Middle Eastern men (*Ilta-Lehti*, 6th April 2008). In general, Finland's Mainstream Media (MSM) would report that these horrific crimes were conducted, as national broadcaster journalist Päivi Annala put it on 3rd December 2018, by men of 'foreign background,' as if Oulu girls had to be worried about being sexually assaulted by Canadian computer programmers or Japanese research scientists at the city's university (Annala, 3rd December 2018).

Long before December 2018, however, everyone in Finland generally inferred that 'foreign background' meant 'Muslim,' especially if the crime in question was rape. And they were right to do so, as data from 2016 found that 93% of all rapes in Finland committed by refugees were committed by those from Islamic countries (*Voice of Europe*, 14th May 2018). The flagrant dishonesty of the Finnish MSM had meant that, by 2017, the previously sky-high trust levels in the homogenous, tiny gene-pooled country of just 5.5 million – its gene pool is both tiny and isolated, meaning Finns suffer from genetic diseases which are rare in the rest of Europe but don't suffer from others which a minority in the rest of Europe do (Kääriäinen et al., 2017) - were coming under strain. In Oulu, an unlikely pair had managed to establish themselves on *You Tube* as the key source of the real news about the city. They live-streamed events, hosted phone-ins, and had contacts with the 'Alt Right' worldwide, such as American psychologist Prof Kevin MacDonald and Dr Jared Taylor, both of whom had appeared on their internet shows. This couple are certainly an anomaly. Junes Lokka (b. 1979) is half-Finnish and half-Moroccan and dresses like a hipster. His fiancée, Tiina Wiik (b. 1985), is a petite former-waitress from the village of Ii, 40 minutes north of Oulu, who speaks inexplicably good English and blogs as 'Swan of Tuonela.' Noticeably short of stature, she is half-Saami; the indigenous reindeer-herding people of northern Lapland who were historically believed, along with the Finns, to be of East Asian origin (Kent, 2014). But, together, they had managed to create an underground movement and, in 2017, Lokka had been elected to the city council, with Wiik as his 'vice-councillor,' standing in for him when he is indisposed (*Oikeusministeriö*, 3rd May 2017).

So it was that on 30th November, Lokka was contacted by the concerned father. This father and a friend of his, who was also the father of a teenager, had set a trap for the Muslim molester in a local park and they'd called the police so that he'd effectively be caught in, or almost in, the act. But, amazingly, one of the dads told Lokka, the police didn't seem particularly interested. Although they arrested the diversifier, they didn't even bother to check the man's phone nor search his flat (*Voice of Europe*, 11th December 2018). If anything, they appeared irritated by what the fathers had done. But, wanting to report the 'real news' to his growing band of followers, Lokka checked the records of who had gone before the courts and on what charges. Though the Finnish MSM and police communications do not provide the names of the accused or even of the convicted, the courts are obliged to. Lokka read through the cases and had an epiphany: 'Rahmani Gheibali (rape), Yosefi Shiraqa (aggravated sexual abuse of a child), Mirzad Javad (aggravated sexual abuse of a child), Barhum Abdullhadi (rape) . . .'

All of these young Muslim men were up before the city's magistrates *at the same time* for rape and sexually assaulting children between June and November 2018. On 16th November alone, one Muslim enricher forced two girls, one 13 and the other 14, to fellate him, while in a separate incident, another under age girl was raped. On 28th November, all these men were in court. It could only mean one thing. Oulu's multicultural enrichment had now gone so far that a line had been crossed. The grooming and rape of underage white girls had arrived in this once quiet corner of North-eastern Europe . . . and it likely went much further than the seven men whom Lokka had found in the court's records. In Rotherham, a post-industrial town in northern England, Muslim grooming gangs had come to light in 2011, after a decade of their activities being ignored by Politically Correct police, civil servants, and social workers (Dearden, 17th July 2017). A group of men, all of them Pakistani Muslims, had been grooming and sexually exploiting white working class girls, aged between 11 and 16, for years, passing them round like sexualised sweeties; taking full advantage of their impoverished and chaotic backgrounds (*The Times,* 5th January 2011). Britain's 'far right' had been warning about them for a very long time, but had been dismissed as paranoid 'racists' – that silencing, indefinite, manipulative term, akin to 'witch' – and even prosecuted, unsuccessfully due to England's jury system (Staff and Agencies, 10th November 2006), for their trouble (Press Association, 17th January 2006). Now, realised Lokka, just as he had long feared, the phenomenon had come to Oulu. And if he didn't act, it would continue to be cloaked in secrecy, with more and more Oulu girls being irreparably hurt.

3. The Muslim Grooming Capital of Northern Scandinavia

So Lokka began to report what was happening online. He published the court list. Word spread and outrage grew. The police were compelled to publically admit that there were seven men suspected of having repeatedly raped or sexually abused young girls, with an eighth man being the one accused of attempting to woo the daughter and stepdaughter of Lokka's contact (*Voice of Europe,* 11th December 2018). *Kaleva* would definitely have been emailed, and would have read, the court records, every day. As such, their journalists knew that a group of Muslim men were in court on these vile charges on 28th November, but they chose not to report it at all, despite routinely reporting many different kinds of crime. In other words, they colluded in hiding what was happening from their readers. Under pressure, on 1st December, *Kaleva* was still so craven – or ideologically fanatical – that it was reporting that a group of 'men' of 'foreign background,' aged between 20 and 30, were accused of sexual crimes against children (Leinonen, 1st December 2018). But within only 2 days, the scandal had fully erupted.

Alternative newspapers, such as *Ålands Nyheter*, based on the autonomous Swedish-speaking island of Åland, reported the full details, including in English so that more people would know: that in mid-November Oulu police had actually arrested seven men for sexual crimes against underage girls and that the suspects were all Middle Eastern Muslims. The girls were aged between 10 and 15 (*Ålands Nyheter*, 4th December 2018). On 4th December, the chief of Oulu police was forced to admit that many of the suspects were 'refugees' – though he didn't specify where from (Julku, 4th December 2018). On the same day, Finland's Interior Minister, half-Russian Kai Mykkänen, publicly condemned what had happened and promised to bring in tougher laws in order to deport foreign criminals, but there was still no mention of the origin of the rapists. He admitted that 'foreigners' were massively overrepresented among rapists in Finland, but failed to clarify that he didn't mean 'foreigners' from England or Estonia (Grasten, 4th December 2018). In February 2018, Mykkänen – Interior Minister despite being then only 37 years old - had asserted that Finland could happily take 10,000 (Muslim) 'refugees' annually, rather than the current 1,000 (*MTV,* 10th February 2018). On 5th December, *Kaleva,* perhaps fearing losing the confidence of Oulu people, inched ever so slightly closer to admitting the truth. Rather than refer to the rapists as 'foreigners,' they specifically used the Finnish word for 'immigrants', a word which everyone understands to, in reality, mean 'non-white immigrants' (Uusitalo, 5th December 2018).

The same day, the Finnish Prime Minister, a millionaire called Juha Sipilä who had studied at Oulu University and hails from nearby Kempele, tweeted his displeasure about events in the northern city. 'The offences in Oulu are appalling. Sexual assault on a child is an inhumane act of inconceivable evil' he wrote (Ålands Nyheter, 5th December 2018). The irony of his condemnation was lost on nobody. Juha Sipilä was elected premier in 2015 on a pledge to let in fewer refugees. However, he almost immediately gave refuge to 32,478 of them in that year alone, due to the supposedly unprecedented nature of the 'refugee crisis' that summer. Sipilä even persuaded Finns to take these Muslim young men into their own houses and promised to do so himself (Withnall, 6th September 2015). Of course, he never did. Under his watch, 'refugees' made their way to Sweden, heard that some benefits were better in Finland and so crossed the Lapland border between the two countries in droves. Evidently in search of 'the first safe country,' many then decided that they disliked Finland and wanted to go back to Sweden, the country's colonial master until 1809 (Kauranen, 25th September 2015). Indeed, in September 2017, a group of 55 male asylum seekers held a protest against Finland's asylum policies in the Finnish border town of Tornio before walking across the bridge to Haparanda in Sweden (YLE, 5th September 2017). But returning to September 2015, less than a month after their arrival in northern Finland, 60 of these desperate asylum seekers held an evening protest in front of Oulu's police station because they didn't think the traditional Finnish food which the refugee centre gave them tasted very nice (Finland Times, 1st October 2015). These protests worked and the menus were changed (China.org, 2nd October 2015). Within a mere two months of the refugees' arrival in northern Finland, these enrichers had gang-raped a 15-year old girl who was walking home from school in Sipilä's home town of Kempele, where some had been housed (YLE, 24th November 2015). The rapists who were before the courts in December 2018 had all been part of the wave that had arrived since 2015. In August 2017, a young Moroccan who had arrive in Finland the previous year as an asylum seeker stabbed two random people to death in the centre of Turku, in the south west.

The authorities described this as Finland's first ever 'terrorist attack,' as it was motivated by fundamentalist Islam (*YLE*, 15th June 2018). When a Somali male walked into the Tuisku Pub in the Tuira district of Oulu in May 2015 and killed two men with an axe, this may also have been terrorism. But this will never be known for sure, because when the police arrived the Somali attacked them as well, so they shot him dead (Hietala, 11th May 2015).

4. Seething Suomi

By December 2018, Oulu was in shock, not so much that there were Muslim rapes – they'd had to deal with that aspect of diversity since at least 2005 – but that these could be systematically organised, directed against underage girls and covered-up by the authorities. The situation developed at breakneck speed, while the Finnish MSM competed to glibly virtue-signal their faux-horror. On the night of 6th December, the 101st anniversary of Finnish independence from Russia, there was a torch-lit parade of nationalists through Oulu. There had been no such thing in previous years. They made their way to the city's large cemetery where, watched over by police, I witnessed them give tense, angry speeches to the be-gloved applause of a largely working class, cigarette smoking audience. On the late afternoon of 10th December, roughly 100 furious Finns braved the cold, sleet and an intimidatingly heavy police presence to protest against the cover up, and the rapes, in front of Oulu city hall. In thirteen years of living in Finland, I had never witnessed anything like it. Finns are, as I already mentioned, stereotypically trusting, cooperative, taciturn and desperate not to offend. Yet, as councillors walked in to the town hall that dark evening, they were greeted with screams of 'traitor!' A working class grandmother from Kokkola – two and a half hours south - ascended the hall's steps, in tears, mourning the death of the Finland in which she grew up: the safe, trusting, Finnish Finland which she had naively assumed would be preserved for her granddaughters. At one point, an innocent, or very brave, Muslim construction worker arose and began to defend Islam from the criticisms which some speakers were vociferously levelling against it. Hissed and jeered by the furious Finns, he was dragged from the steps and beaten up, with the police marching him away, not daring to inflame the mob by arresting those who'd slapped him around until he squealed. It was revealed that many of the rape cases had centred around Oulu's monstrosity of a new shopping mall, Valkea ('White'). This shopping centre had become a Mecca for both Finnish teenagers and Muslim males, who would hang around in it day and night, trying to chat up under age Finnish girls in English or broken Finnish.

Two weeks later, on 26th December, it was reported that the windows of Oulu's small mosque had been smashed. The head priests, known as *Kirkkoherrat,* of the city's assorted Lutheran 'congregations' wasted no time in sending out a press release virtue-signalling how appalled they were by this (*Kaleva,* 26th December 2018). They conveniently forgot to mention that the mosque's imam, a Bangladeshi scientist called Dr Abdul Mannan, was a fundamentalist Muslim who began the Acknowledgements section of his doctoral thesis on geology with the words: 'Thanks to God and may His peace and blessings be upon all the prophets for granting me the chance and ability to successfully complete this study' (Mannan, 2002). In 2008, I was working for an English-language online newspaper aimed at foreigners in Oulu called *65 Degrees North*. I interviewed Dr Mannan for an article about his mosque. A friendly and welcoming chap, Dr Mannan informed me that genuine Islam involves Sharia Law and that amputating the hands of certain classes of criminals is part of Sharia Law and is Allah's will (Dutton, 26th March 2008). In 2009, I conducted participant observation fieldwork at the city's mosque. One evening, at my Finnish language course, I got talking to a Tajik doctor about this mosque. He told me that he would never attend such a mosque, because it was 'Wahabist' (see Nahouza, 2018) – this being the form of strict Sharia Islam practiced in Saudi Arabia (Dutton, 2009a, p.32). Indeed, a Bangladeshi called Taz Rahman moved to Finland, 'met Dr Abdul Mannan, imam of a mosque in Oulu, Finland. He became influenced by Mannan's radical religious views,' married Mannan's daughter, and was killed in 2017, fighting for ISIS in Iraq (Labu, 18th May 2017). This was of particular interest in Finland, at least in the alternative media, because Mannan was not just Oulu's imam but a Social Democrat vice-councillor (*MV??!! Media,* 12th May 2017).

In early January 2019, more grooming cases were revealed meaning that 16 Oulu Muslims were now suspects (Malm, 14th January 2019). Poignantly, it also came to light that in August 2018, a 14-year-old girl had been persuaded, by Muslim men, to leave Valkea and go to their flat to get drunk with them. They raped her. In

October, her 12-year-old sister went into the girl's bedroom to find that she'd hanged herself with a belt. Word spread online and Oulu's police took to the newspapers to deny this story. But a week later, on 11th January, they were forced to admit that, in effect, Sipilä's 'refugees' had, indirectly, claimed their first Oulu life (*Suomen Uutiset,* 11th January 2019). The next day, roughly 50 members of the anti-Muslim vigilante group Soldiers of Odin rallied at Valkea – now heavily police-patrolled and clear of Muslim men – and marched through the snow blanketed city. In the evening, to quell the public mood, both the Prime Minister and the Interior Minister gave press conferences about the situation at Oulu city hall where MSM journalists asked them easy questions. I accompanied Junes Lokka, as I was to be interviewed on Tiina Wiik's internet show, *Happy Homelands,* later that evening. Junes suggested to them that there was nothing unique about Oulu which explained why these things had happened other than the presence of large numbers of Muslim men. And the next day, lo and behold, cases of Muslim grooming were reported in Helsinki. On 13th January, the British media personality Katie Hopkins came to Oulu to look into what was going on. The police and mayor refused to talk to her and she and Councillors Lokka and Wiik got themselves kicked out of the city's refugee reception centre (Honkanen, 15th January 2019).

By mid-February, further incidents had come to light. One incident took place in a flat on 17th November with *YLE* reporting: 'One of the victims is under 16 and the other under 18 . . . The suspect has a foreign background and has a Finnish residence permit' (*YLE,* 12th February 2019). Another sexual assault took place in January in the run-down, immigrant heavy suburb of Tuira, the victim being a 14-year-old girl (*YLE,* 13th February 2019). Even this deep into the crisis YLE were censoring the religion and ethnic origins of the perpetrators. On 20th February 2018, police commissioner Markus Kiiskanen announced that the police were investigating 29 men, most of them 'foreign,' for child rape and sexual assault in Oulu, with one of the victims being 10 years-old. Kiiskanen failed to give precise details about what kind of 'foreigners' these men were and nor did the MSM inquire (*MTV,* 20th February 2018). By 2nd March, Lokka and Wiik were in Helsinki to meet the British political activist Tommy Robinson (*Suomen Uutiset,* 2nd March 2019), founder of a street activism group called the English Defence League. Robinson, who had recently been in prison for reporting outside a trial of Muslim child groomers in Oxford. He was fascinated by the way in which 'citizen journalism' had uncovered Islamic grooming in Finland just as his citizen journalism had in the UK.

5. Why Oulu? Why Finland?

But despite his role in these events, Juha Sipilä did at least raise an interesting question. Why was this happening in Oulu? And, more generally, why was it happening in Finland? This 'Oulu in northern Finland' grooming scandal had even made the international press, having been reported, for example, in Britain's *Daily Mail,* for which Katie Hopkins use to write, before being forced out for defaming someone (Malm, 14th January 2019) as well as in *The Sun,* the UK's most popular newspaper (Hodge, 6th December 2018). This specific problem wasn't being reported in other Nordic countries; countries which had experienced Muslim immigration since the 1970s. So why was it happening in Finland? Why was it that Oulu had changed so rapidly since 2003?

In this book, I will argue that the central reason is relatively simple, though its causes are complex and reach deep into the Finns' evolutionary history: 'Finns are too nice.' I will show that Finns are the most intelligent people in Europe and that intelligence predicts kindness and low self-esteem, meaning a strong desire to conform. I will demonstrate that the country's freezing, inhospitable ecology has selected for a small gene pool and highly cooperative – 'nice' – groups of people. This means that there are very few high IQ outliers, helping to explain the country's conformity and also, as we will see, it's 'niceness.' I will show that the country's modal personality – the stereotypical 'Silent Finns' – reflects this ecology in the same way. People are extremely high in the personality traits of altruism and Conscientiousness and the small gene pool – harsh ecologies lead to small gene pools, because you must be strongly adapted to them or perish – means there are very few outliers, who might 'rock the boat' and question a given situation. I will demonstrate that originality stems from outlier high IQ antisocial people and that Finland simply has fewer of these, per capita, than do other European countries.

This has serious consequences. People who have low self-esteem, and are desperate to please, will imitate and fail to question those whom they look up to. Those who are too trusting will be too trusting of their leaders, meaning that democracy will become debased far more easily and bad leaders won't be deposed until it's too late. Ethnic groups like this will lack a class of people who will be prepared to question the *status quo*, and particularly the actions of the society's leaders. And, ultimately, very high empathy is associated with schizotypal (schizophrenia spectrum) traits, meaning a higher percentage of a 'nice' population will tend towards extreme social anxiety and even a tendency to not perceive the world as real, resulting in apathy in the face of immediate peril. I will show that it is Finland's unusual national psychology – grounded in centuries of evolutionary adaptation to an extremely cold and harsh ecology – that helps to explain why it has been so rapidly enriched with Islamic child rape gangs. And I will demonstrate that this adaptation means that the environmental processes set off by industrialization – such as the decline in religiousness – will have even more severe maladaptive consequences for the Finns than they will for other ethnic groups, because the evolutionary situation for Finns has been so precarious. It is for all these reasons that Finns are being raped and we will look at why, from an evolutionary perspective, some of Finland's Muslim refugees engage in rape.

However, we will also see that, due to this ecology, Finns can be persuaded, in a way that is less true of other ethnic groups in Europe, to make extreme sacrifices for their group and to act with incredible brutality when threatened by an out-group, at least under conditions of relative poverty. It is this adaptation that Finns need to rediscover if they wish to save Oulu and Finland itself. And I will argue that their evolved predisposition towards ethnocentrism combined with their strong conformism means as Multiculturalism leads to societal collapse, Finns are likely to return to nationalism, and begin the process of reviving themselves, more quickly than other Western European nations.

Chapter Two

Understanding Intelligence

1. What is Intelligence?

Before we can begin to make sense of why this process has happened to Finns so quickly, there are two key concepts which we have to understand: 'Intelligence' and 'Personality.' This is because it is the evolution of these in a certain extreme direction, among Finns, that has played such a key role in what they have allowed to be done to themselves. Let us begin with the more contentious of the two 'concepts': 'intelligence.'

'Intelligence' is defined as 'the ability to reason, plan, solve problems, think abstractly, comprehend complex ideas, learn quickly, and learn from experience' (Gottfredson, 1997, p.13). In other words, it is the ability to solve complex problems and to solve them quickly. The quicker you can solve the problem then the more intelligent you are and the harder the problem has to be before it is simply beyond you then the more intelligent you are. Intelligence is quantified by the intelligence quotient (IQ), a score that can be derived from a number of standardized tests. These tests measure reasoning ability across a wide range of areas, for example verbal, mathematical, and spatial. Ability in each of these subsections positively correlates with ability in the other subsections (see Lynn, 2015), allowing us to posit that they together measure an underlying factor called 'general intelligence' (g). This is the essence of intelligence, and it is generally what we are talking about when we say that one person is more 'intelligent' than another. A correlation, it should be noted, refers to the relationship between two variables; the extent to which one predicts the other. This can be either a positive or negative relationship. A correlation of 1 means that one variable perfectly predicts the other. 'Statistical significance' is how scientists test, using calculations based on the strength of the correlation and the size of the sample, whether or not the correlation is merely a fluke. It is accepted, based on this, that if we can be at least 95% certain it is not a fluke then the relationship is 'statistically significant' ($p = <0.05$) and thus real.

It cannot possibly be argued that intelligence is somehow 'a very Western concept,' something which a female Finnish cultural anthropologist once told me, with great profundity, at a Halloween Party in 2014. Proxies for intelligence include general knowledge, something which is valued in all cultures (Buss, 1989). Intelligence is negatively associated with criminality, which is disliked in all cultures. It is also robustly correlated with education, income, and health. As such, intelligence cannot be dismissed as context-dependent or as only relevant in the West and nor can it be dismissed as unimportant. A full list of the traits associated with intelligence can be seen in Table 1. It is worth noting at this stage – and we'll look in more detail at this later – that many behaviors stereotypically associated with the Finns are also correlated with high intelligence but that others, such as alcoholism, are clearly anomalous.

Table 1. Behaviors and Preferences Associated with Intelligence (Dutton & Woodley of Menie, 2018; developed from Jensen, 1998).

Positive Correlation	Negative Correlation
Achievement motivation	Accident proneness
Altruism	Acquiescence
Analytic style	Aging quickly
Abstract thinking	Alcoholism
Artistic preference and ability	Authoritarianism
Atheism	Conservatism (of social views)
Craftwork	Crime
Creativity	Delinquency
Diet (healthy)	Dogmatism
Democratic participation (voting, petitions)	Falsification ('Lie' scores)
	Hysteria (versus other neuroses)
Educational attainment	Illegitimacy
Eminence and genius	Impulsivity
Emotional sensitivity	Infant mortality
Extra-curricular attainments	Obesity
Field-independence	Racial prejudice
Height	Reaction times
Health, fitness, longevity	Religiousness
Humour, sense of	Self-Esteem
Income	Smoking
Interests, depth and breadth of	Single/young motherhood
Involvement in school activities	Truancy
Leadership	Trust (lack of)

Linguistic abilities (including spelling) Logical abilities	Weight/height ratio (BMI)
Marital partner, choice of	
Media preferences	
Memory	
Migration (voluntary)	
Military rank	
Moral reasoning and development	
Motor skills	
Musical preferences and abilities	
Myopia	
Occupational status	
Occupational success	
Perceptual abilities	
Piaget-type abilities	
Practical knowledge	
Psychotherapy, response to	
Reading ability	
Social skills	
Socioeconomic status of origin	
Socioeconomic status achieved	
Sports participation at university	
Supermarket shopping ability	
Talking speed	
Trusting nature	

2. What are IQ Tests?

Intelligence is measured by IQ tests. We know that these are reliable because their results correlate with other intuitive measures of cognitive ability, such as educational success. . IQ is defined as having a population mean of 100 and a standard deviation (SD) of 15, which means that your intelligence is quantified relative to others from the same population group, which is typically those who live in your country. Intelligence increases with age during childhood, but has usually reached its phenotypical level in early to late adolescence. Even a below average 16-year-old performs better than a very clever 3-year-old who can already read. IQ is therefore calculated relative to people the same age. The average person scores 100; anything less than this is below average, and anything above it is above average. Rather like height, IQ is 'normally distributed' on what looks like a bell curve. Most people have an IQ of around 100 with the percentages with lower and higher IQs tapering off on both sides. Because one standard deviation (SD) is already defined as 15 points, 68% of people have an IQ between 85 and 115, and 95% of people have an IQ between 70 and 130 (±2 SD). This is the 'normal' range. If you score below it, you are classified as retarded. If you score above it, as the kind of people reading this book probably do, then you are exceptionally bright.

But what about specific cognitive abilities? We all know people who have very good language comprehension but poor mathematical ability. There are certainly such individuals, but across a large group of people it turns out that all specific cognitive abilities are positively correlated with each other, and that this is also true of most specific cognitive abilities for most individuals. This fact is what lays the foundation for the g factor (general intelligence). The validity and reliability of IQ tests has been subject to criticism, and even the concept of intelligence has been questioned. A very popular alternative was proposed by American educational psychologist Howard Gardner (1983), arguing that there are 'multiple intelligences,' such as bodily-kinesthetic, emotional, musical, and interpersonal. However, the thrust of Gardner's model is that these abilities are independent, as they are realized by different parts of the brain, and that a person can excel in one or more 'intelligences 'while being below average in others. That is, however, not what reality looks like. Higher IQ people perform better in most cognitive domains, and lower IQ people perform worse. Gardner's assorted kinds of 'intelligence' are either a misuse of the word 'intelligence' or they are simply examples of cognitive abilities underpinned by 'intelligence' as normally defined. Emotional intelligence is a concept that has also been launched as a competitor to IQ or general intelligence. This ability to deal with other people and solve social problems is, however, weakly positively correlated with IQ at about 0.3 (Kaufman et al., 2011). The more intelligent you are then the more empathetic you are; the better you can imagine what it is like to be someone else.

Criticisms have also been leveled against IQ tests but none of them really work, despite the great profundity with which they are often expressed. IQ tests have been found to have high predictive validity for school achievement (and, thus, other measures of cognitive ability), occupational status and criminality (negatively) (Jensen, 1998). They cannot be argued to be substantially culturally biased, as they correlate with objective measures, such as reaction times (negatively) and cranial capacity (positively): they correlate with how quick you are to respond to stimuli and simply how big your brain is; unsurprisingly, the brain being a thinking muscle. The robust negative correlation with reaction times (see Table 1) – how quickly you react to a stimulus, such as a light being switched on; the cleverer you are the shorter your reaction time - implies that intelligence can substantially be explained by a high functioning nervous system (Jensen, 1998).

One very common argument against the validity of IQ tests is the fashionable concept of 'stereotype threat.' The idea is that people who belong to a group believed to do badly on IQ tests – such as black people - do badly because of their own expectations that they will do badly, which presumably stresses them out, reducing their performance. However, meta-analyses have found that this effect is mostly non-existent, and when it can be found – for certain people in certain situations – it is much smaller than the group differences it is used to explain away. Some studies have found that groups told they will do badly on IQ tests because of some factor about them seem to consequently do better than expected and there is a huge problem of publication bias, in this area with studies disproving stereotype threat simply not being published (Ganley, 2013). Also, the theory does not at all address the critical question of how systematically incorrect stereotypes could originate, as 75% of racial stereotypes have been found to be at least partly accurate and 50% completely accurate (Helmreich, 1982). The simplest explanation, as has been empirically explored in detail by American psychologist Lee Jussim (2012), is that stereotypes develop because they are broadly true.

3. Are there Genetic Race Differences in Intelligence?

The possibility of race differences in intelligence is one of the reasons why the entire area of research is so 'controversial.' Indeed, incredibly, attempts were made, in 2004, to prosecute Finnish political scientist Prof Tatu Vanhanen (1929-2015) when he told *Helsingin Sanomat's* magazine that his academic research had

165 cm = 5'5"   185 cm = 6'

unearthed precisely such differences (Dutton, 2015a). A 'race' is a breeding population separated from other such populations for sufficient time to develop distinct sets of gene frequencies, tending to express themselves in physical and mental differences which are adaptations to their different environments. Geneticists have highlighted around 10 highly distinct genetic clusters, which largely coincide with different 'races' (see Lynn, 2015). This is useful as, precisely because they differ in gene frequencies, 'race' allows important predictions to be made about modal intelligence, personality, the frequency of certain genetic and partly genetic medical conditions, response to drugs and drug dosage, blood type and much else (see Dutton, In Press). Also, the argument that race differences are miniscule and that there are 'more differences within races than between them' makes absolutely no sense. The fact that there is more variation in height within each sex does not invalidate the observation that men are on average taller than women. A 20 cm range characterizes most men (165-185cm) and most women (150-170cm), and yet this 15 cm mean difference is quite sufficient to predict the ability to reach a tall shelf or to enter a confined space. Similarly, there would be tiny genetic differences between a bog standard musician and a Mozart, but these tiny differences, when in the same direction, come together to snowball into huge consequences. There is only a small genetic difference between chimpanzees and humans, but separating them into different species allows successful predictions to be made, this being the whole point of scientific categorization. Further, if a number of small differences all push in the same direction – because they are adaptations to a specific ecology – then this will result in clear and predictable racial differences, so it is irrelevant that there may otherwise be a high level of diversity within races (Cochran & Harpending, 2009). A final crucial point is that genetic predispositions cause individuals to choose different environments, preoccupations, and people to interact with, which leads them to develop certain abilities and skills more than others, so called gene-environment interaction.

As for race and intelligence, a meta-analysis by Richard Lynn (2011a, p.101) based on twin studies found that intelligence had a heritability, on adult samples, of 0.83. It might be argued that intelligence is highly heritable but that environmental factors explain, for example, why US blacks have lower IQ than US whites. This seems most unlikely, however and the reasons for this have

31

been set out by the American philosopher Michael Levin (2005) in his book *Why Race Matters*: The 15-point difference between white and black IQ scores in the USA is evident by the age of three. The earlier a difference becomes evident, it is argued, the more likely it is to be genetic (Broman et al., 1987). Interracial adoption studies have shown that black adopted children's adult IQ has no relation to their white adoptive parents' IQ, but it is very similar to that of their biological parents (Weinberg et al., 1992). The more resistant a difference is to interventions the more likely it is to be genetic. British-Canadian psychologist J. Philippe Rushton (1943-2012) (Rushton, 1995) has noted that black people have been argued for millennia, even by Moorish explorers, to have low average intelligence and that attempts to boost their intelligence, based on environmentalist assumptions, have had no significant impact. Moreover, IQ tests can be divided into different subtests and ability in these subtests varies in the degree to which it is genetic. Blacks score the worst on the most 'culture fair,' which are also the most strongly genetic, parts of the IQ test (Jensen, 1998). And most importantly, moving beyond Levin's discussion, Italian anthropologist Davide Piffer (2016) has found genetic evidence for average national IQs. The average frequency in the population of genetic variants which are correlated with extremely high educational attainment, something which is very strongly associated with high IQ, is correlated at 0.9 with national IQ. In another study, Piffer (2018) replicated this finding with a sample of 1.1 million people from 52 countries. In effect, Piffer provides extremely persuasive evidence that race differences in intelligence are overwhelmingly a reflection of genetic differences. Critics are left with one final argument, which is that the hypothesis fits the evidence but it *must not* be true or must be kept secret because it is 'dangerous,' or because it might be 'offensive.' It can be responded that this is a fallacy: an appeal to consequences. That it may be dangerous or offensive is irrelevant to whether or not it is true. Moreover, there are numerous serious potential dangers to building public policy around inaccurate information.

That there are genetic race differences in IQ should be no more offensive to any non-white person than the degree of offence I might feel if an Ashkenazi Jew asserted that Ashkenazi Jews have a higher IQ than white people. I don't feel any offence. It's a fact. Ashkenazi Jews in the USA have an average IQ of 112 (Lynn, 2015). In much the same way, it's a fact that university graduates have a higher mean IQ than non-graduates. It doesn't make any sense that all non-graduates should harbour pain and resentment over the fact that they, as a group, have lower intelligence than graduates. As is always the case, variation within any group is large enough that the group average does not determine the psychological properties of any particular individual. I can testify, from having attended Durham University in England and then Aberdeen University in Scotland and from teaching in the Cultural Anthropology Department at Oulu University, that you can get some very stupid university graduates and some very clever non-graduates. You are unlikely to find a mentally retarded university graduate or, these days at least, a mathematical genius who has never gone to university. What is true of the group mean has only a limited bearing on what a member of the group will be like. But it still means that life with a more intelligent group of people, such as the Finns, will be noticeably different from life with a group who are less intelligent than they are.

Chapter Three

Personality and Life History Strategy

1. What is Personality?

Personality is defined as 'the combination of characteristics or qualities that form an individual's distinctive character.' Thus, personality can be seen as a set of variable traits (McAdams & Pals, 2006, p.212). A number of personality models have been developed and are used in parallel but in recent decades the so-called Big Five model has come to dominate. The Big 5 is a bottom-up model, based on actual correlations between people's ratings of a large pool of statements about themselves (Nettle, 2007). Various personality traits, such as 'warmth' or 'depression', are been found to correlate positively or negatively with each other, but to have no correlation, or only a very weak correlation, with other personality traits. Other models are more top-down, based on inductive and deductive reasoning about what traits should be relevant in order to be adaptive to different environments and situations. Fortunately, we do not have to choose between bottom-up and top-down, or even amongst specific models, because all the more influential models tend to exhibit the same traits, although their names might differ. Regardless of personality model, therefore, twin studies show that the heritability of personality traits are in the region of at least 50% (Nettle, 2007) and possibly up to around 66% (Lynn, 2011a). For the present discussion, the Big Five serves at least as well as any other model, and it is also currently the most widely accepted and these variables are regarded as substantially independent of intelligence. The Big Five are:

1. *Extraversion*: Those who are outgoing, enthusiastic and active, seek novelty and excitement, and who experience positive emotions strongly. Those who score low on this express Introversion and are aloof, quiet, independent, cautious, and enjoy being alone.

2. *Neuroticism*: Those who are prone to stress, worry, and negative emotions and who require order. The opposite is those who are Emotionally Stable and who are better at taking risks.

3. *Conscientiousness*: Organized, directed, hardworking, but controlling. The opposite types are spontaneous, careless, and prone to addiction.

4. *Agreeableness*: Trusting, cooperative, altruistic, and slow to anger. This is contrasted with those who are uncooperative and hostile. There are two key aspects to Agreeableness: altruism and empathy (theory of mind; the ability to intuit what others are thinking). These positively correlate but they are not the same thing.

5. *Openness-Intellect*: Those who are creative, imaginative, aesthetic, artistic, and open to new ideas (Nettle, 2007). This is contrasted with those who are practical, conventional, and less open to new ideas. The traits which compose Openness such as 'unusual thought patterns,' 'impulsive non-conformity,' or 'aestheticism,' are often only weakly correlated. This has led some researchers to suggest that Openness-Intellect is, essentially, a combination of intelligence (intellectual curiosity), low Conscientiousness and low Agreeableness, and should simply be abandoned (e.g. Dutton & Charlton, 2015)

In each case, the traits are conceived of as a spectrum and are named after one extreme on the spectrum. They are considered useful because variation in the Big Five allows successful Life History predictions to be made. For example, the Termanites were a cohort of 1,500 Americans of above average intelligence first surveyed in 1921 and then finally in 1991. Drawing upon them, it was found that extraversion, independent of any other factor, was a predictor of early death. It increased the risk threefold, probably because Extraversion encourages you to take risks in pursuit of the high positive emotional reward (Friedman et al., 1993). Low Agreeableness predicts criminality and divorce. High Conscientiousness predicts doing well in the worlds of education and work, while low Conscientiousness predicts criminality, low socioeconomic status, and addiction. High Neuroticism predicts depression, anxiety, marital breakdown and aspects of creativity (Nettle, 2007). It is also associated with being a religious seeker and having periodic phases of religious fervour (Hills et al., 2004) as well as, very specifically, with heart disease (Cukic & Bates, 2015). An optimally high level of Openness is associated with artistic success. (Nettle, 2007). These personality characteristics also predict the kind of subjects that will interest you. Science types tend to be high in Conscientiousness, Agreeableness and Intellect, and low in Neuroticism and Openness. In the Humanities this pattern is reversed (De Fruyt & Mervielde, 1996).

As already noted, the Big 5 are substantially independent of each other, though there is a correlation at the level of the aspects of which they are composed. Specifically, what we might call the socially positive aspects of each trait do in fact correlate. These are the aspects which make you a socially effective person – friendly, diligent, cooperative, reliable, open-minded – meaning, in essence, that you 'get on in life.' As such, personality can be reduced down to a 'General Factor of Personality,' much as intelligence can be reduced down to the *g*-factor. People can be positioned higher or lower on a spectrum measuring this General Factor of Personality, GFP for short. GFP is associated with socioeconomic success (Van der Linden et al., 2010).

2. National Differences in Personality Proxies

If a person has an extreme placing on any of the Big 5, or is very low in GFP, then they are understood to have a 'personality disorder,' in other words they are 'pathological.' Usually, these disorders do not directly cross-over with the Big 5. A personality disorder is characterized by maladaptive patterns of behaviour and inner experience which cause the sufferer to differ sufficiently from the norms of their own culture so as to be regarded as pathological. The symptoms tend to cluster together, meaning that psychologists have been able to identify a number of distinct personality disorders. Some people are more pronounced or clearer exemplars of these disorders than others. Equally, many disorders involve extremes of personality, meaning we can conceive of a spectrum, with 'normal' in the middle. National differences in these kinds of pathologies are highly relevant, because it is notoriously difficult to obtain sound data on national differences in personality traits, because the traits are self-rated, and hence to a considerable extent relative to one's own population. Comparing self-ratings across countries does not, therefore, establish that Finns are, for example, higher in Conscientiousness than the British, although this is almost certainly the case. On so many occasions, I have watched Finns wait at zebra crossings, because there is a red man, even though there are clearly no cars coming. British people, in such circumstances, would simply cross the road. But Finns only seem to pluck up the courage to do so – and 'break the rules' – if I lead the way.

However, this difference almost certainly wouldn't be recorded if we were to compare representative samples of Finns and Brits on scores on some test of the Big 5. If anything, the British would probably score higher in Conscientiousness. Consistent with this, international comparisons have produced intuitively bizarre results, such as that the Japanese are very low in Conscientiousness and that Sub-Saharan Africans are very high in it (Schmitt et al., 2007). This is because the personality surveys involve people subjectively evaluating the degree to which they are 'tidy,' for example. But they will be making this evaluation according to different cultural norms of how tidy 'tidy' really is, meaning it is extremely problematic to compare different cultures in this way (see Meisenberg, 2015). Accordingly, ascertaining the percentage of a population who suffer from a pathology – which can be measured more objectively – better allows us to compare populations on average personality, because the pathology will tend to be an extreme form of certain personality traits, potentially reflecting a difference in the average on these traits. There are many such trait spectrums, but the most relevant one for our purposes is what we might call the 'empathy' spectrum.'

Autism is characterized, in part, by poor theory of mind and low empathy. Low empathy involves the inability to 'mentalize' and so imagine the feelings of others or read mental states from environmental cues. An autistic will not understand how someone feels from subtle facial expressions or mild verbal indications. In this narrow sense, empathy sits at one end of a spectrum. At the other end of the spectrum is schizophrenia. This is characterized by over-mentalizing (Badcock, 2003). Schizophrenics are obsessed with other people's minds, they read far too much into cues, and this makes them paranoid or simply wrong. Thus, a subtle cue indicating that someone is slightly annoyed or upset may be understood by a schizophrenic to mean that the person is dangerously angry and wants to kill them. By contrast, the low empathiser will likely read nothing into it at all. Schizophrenia and low empathy are the extreme ends of this spectrum, but there are various subtler deviations from 'normal' as we move away from these extremes. There is a range of severity to schizophrenia, with mild symptoms summarised as 'schizoid personality.' This is characterised by anhedonia (the inability to feel joy) and apathy. More severe is schizotypal personality, where the schizoid symptoms are accompanied by social anxiety (Hodgekins, 2015, p.184), paranoid ideation, unconventional or paranoid beliefs and derealisation, where you don't quite see reality as 'real.' Diagnosable schizophrenia is a particularly severe manifestation of these characteristics (Dowson & Grounds, 2006). Asperger's is a mild form of Autism where the sufferer has some autism traits or all such traits but only weakly.

3. Testosterone and Genius

Another method to compare population differences in personality is to look at the population prevalence of gene forms – polymorphisms - that are associated with specific extremes of personality. Yet another method is to explore population averages in markers of testosterone. People who are high in testosterone tend to be aggressive, driven, uncooperative and highly sexed. They are low in Agreeableness and Conscientiousness. There are many markers of testosterone for which national-level data is available including left-handedness, hand shape (the so-called 2D:4D ratio, whereby a smaller distance between the longest finger and the shortest finger reflects higher testosterone), hairiness of fingers, prostate cancer prevalence, number of sexual partners, and frequency of sexual intercourse (Van der Linden et al., 2018). Indeed, Autism is associated with markers of high testosterone (Baron-Cohen, 2002) while Schizophrenia is associated with markers of low testosterone (Agarwal, 2013). This should be no surprise. Testosterone is a masculinizing hormone. The stereotypical 'male brain' has been shown to focus, rather like many intelligent autistics do, on systematizing while being poor at empathising. The stereotypical female brain is quite the reverse.

Yet another index of national differences in personality would be per capita geniuses, when controlling for IQ. This would also substantially control for national differences in economic development. A characteristic of genius is to come up with fantastically original ideas. The reason why this is important is probably the highly social structure of human endeavour, which on the one hand helps us build on the achievements of others, but on the other hand lures us into following paths already trodden, which prevents leaps to the novel and fundamentally different. It is widely agreed that the 'genius,' in science at least, is characterized by outlier high IQ combined with moderately low Agreeableness and moderately low Conscientiousness. He is also obsessive and driven. These character traits – in combination with outstandingly high IQ – are crucial. As a person who 'breaks the rules,' he can 'think outside the box' and thus develop something of pronounced originality. But he must also be prepared to present this finding. Original ideas are almost always met with hostility, because they offend against vested interests. But, being low in Agreeableness, the genius won't care about offending people and, in that he'd also be low in empathy, probably couldn't anticipate offending people even if he did care. In other words, a genius is a manifestation of outlier high IQ plus an optimal level of low empathy (see Dutton & Charlton, 2015) or, to put it another way, outlier high IQ plus relatively high testosterone. Consistent with this, Van der Linden and colleagues (2018) found that when you look at countries with an IQ of at least 90 then national testosterone level (based on a range of testosterone proxies) predicts a nation's per capita number of science Nobel Prizes and even citations in important scientific journals. So, per capita genius is an important insight into a nation's average personality. Geniuses are generally born to parents who are highly intelligent but within the 'normal' range. They are products of unlikely but possible genetic combinations, meaning that when they have children their children do not tend to be anywhere near as eminent as they are (Dutton & Woodley of Menie, 2018, Ch. 6).

4. Life History Strategy

Now, everything that we have discussed so far – population differences in intelligence, personality, schizophrenia and and testosterone – can ultimately be reduced down to one beautifully simple means of understanding group differences: Life History Strategy.

A fast Life History Strategy (LHS) is a suite of physical and mental traits that develop in an easy but unstable ecology, in which you must respond to sudden, unpredictable challenges by being extremely aggressive. A fast LHS is called an r-strategy, while a slow LHS is known as a K-strategy (Woodley, 2011a). Those who are r-strategists 'live for the now'. They live fast and die young, because they could be wiped out at any moment. They tend to be born after less gestation but more developed, they reach childhood developmental milestones more quickly, enter puberty earlier, begin their sex life younger, engage in promiscuous and frequent sex in order to procreate as much and as quickly as possible, invest little in their partners or offspring, and they tend to die relatively young, as they age more quickly. As they invest more energy in sex, they need to advertise their genetic quality as conspicuously as possible – because they could be killed with a second's notice – so they tend to grow pronounced secondary sexual characteristics, such as large breasts, in order to display the fact that their genes are good enough to allow them to do so. They create weak social bonds – 'What's the point of bonding? The other person may be suddenly killed so they'll never be a pay off.' The ecology is so favourable that basic needs are pretty much met. So they have poor impulse control and low altruism; allowing them to respond with massive aggression to very sudden dangers. Extraversion is high as, in an unpredictable ecology, you may as well take risks. They also have high mental instability. The easiness of the ecology means there's less of a need to develop cooperative groups to survive, so there's less selection pressure against mental instability.

An extremely slow LHS occurs in a harsh yet stable environment, because this means that the carrying capacity of the ecology for a particular species is reached. Accordingly, its members start to compete much more strongly with each other. They do this via an arms race of adapting to the environment, meaning they have a smaller gene pool. They begin to invest less energy in procreation and more energy in growth, adaptation, and, crucially, the *nurture* of their offspring. In such an ecology, somebody could have numerous children but they might all die of cold and starvation. By contrast, the competitor who invested in nurture would ensure that his fewer offspring would grow up to be adapted to and to understand the harsh yet predictable environment, so his offspring would survive. The result is that life slows down, so people can learn about their complex but predictable ecology. Puberty comes later, as does the beginning of a person's sex life. The slow LH strategist has fewer sex partners, has less sex with them, and invests more resources in them. This selects for higher altruism, as there is competition for mates and you're more likely to survive if you can get along with people; and higher impulse control, allowing you to plan for the future. Being more predictable an environment, these efforts tend to pay off in the long run, and there is more likely to be a payback for cooperation as people live longer and are prone to invest in reciprocity.

In harsh environments, people who are part of cooperative groups are much more likely to survive, as are cooperative groups in general, further elevating these personality qualities, leading to people who create very strong social bonds. Extraversion decreases because the results of risks become easier to predict. People's bodies also change. More energy is invested in developing a complex mind, so less is invested in the body. Also, the mind of a potential spouse becomes more important, and more of an object of attraction, if, as a male, you are going to invest in that spouse. You want to be sure that your offspring are really yours and that the spouse will be a good mother, so you select for Conscientiousness and Agreeableness. This means that there's less of a need to conspicuously advertise gene quality, so breasts, for example, start to become smaller. These high K groups impose strict norms and are highly conformist (Dutton et al., 2016a). At the level of environmental influence on personality, a cooperative personality can only be developed if children's tendencies to be selfish and impulsive are removed; in other words, the spirit of impulsive non-co-operators has to be 'broken.' Thus, as the high K group is so strongly attuned to its environment, its members are more environmentally plastic: more has to be learnt and less is simply instinctive. Hence, K-strategists have longer childhoods. They are more environmentally sensitive than r-strategists; 'culture' – rather than instinct – is more central to their lives (Sng et al., 2017).

As the group becomes more K, its niche becomes more specific, because the harsher and more predictable the ecology is the more specifically adapted you must be to survive. In an easy ecology you can forage for food all year round, but in a harsh one you must specialize, innovating very specific techniques and systems to catch the (relatively rare) sources of food. This means that the different components of K end up being less strongly inter-correlated, because selection favours the highly environmentally *specific*. Paradoxically, an extremely high K group can be quite r-selected in certain specific ways precisely because its ecology is so harsh and predictable (Woodley, 2011a). Rushton (1995) has shown that there are numerous consistent race differences in Life History Strategy, with Sub-Saharan Africans the fastest, Northeast Asians the slowest and Caucasians (Europeans, South Asians, North Africans) intermediate but closer to Northeast Asians. However, Japanese psychologist Kenya Kura and his team have shown that the Japanese are, in essence, extremely *K*-selected and so very high in conformism and social anxiety, concerned about what others think of them (Kura et al., 2015). But they are concomitantly highly r-selected in the sense that they can be very aggressive and hostile towards foreigners (Dutton et al., 2016a) and those whom they regard as impure, leading to very low levels of adoption, despite adoption being – from an evolutionary perspective – nurture for the sake of it and thus a *K* trait (Dutton & Madison, 2018a).

Of course, this implies that there should be not just 'race' differences in LHS but also differences between ethnic groups within races. You've probably guessed by now that Finns are slow Life History Strategists compared to other European ethnic groups. Let us now look at the evidence for this.

Chapter Four

The Cleverest People in Europe

1. Cold Winters Theory

Finland is a pronounced example of a harsh yet predictable ecology. The Finnish winter can be quite fantastically cold. On 23rd December 2010, my parents came from England to Oulu to stay for Christmas. Upon their arrival, we quickly made our way to the local shop to stock-up on beer, due to the inconveniently restrictive times during which you could purchase alcohol in Finland at the time. My father, who accompanied me to the nearby corner shop *Siwa*, had experienced a Finnish winter before, but never anything like this. Failing to heed my warnings of the severity of the situation, he wasn't sufficiently wrapped up and spent most of the journey back from the shop, which was into the wind, holding his nose with his gloved hand. It was -36 °C and, with the wind, it must have felt a lot colder. If you were to stand inside a freezer it would be –17 °C. But this was almost 20 degrees lower than that; the difference between a very cold day in England (0 °C) and a pretty hot one. As I write this, on 30th January 2018, it is -10 °C. I know that because I've just looked at the thermometer outside my kitchen window. Almost all Finnish houses have one of these, because if you don't dress properly for the weather, and you stay outside too long, then you may well die.

Finns know how to dress for the winter and how to keep warm. The windows of houses are not double but triple-glazed, to keep the heat in. Cats lack cat-flaps, because they'd let the heat escape and, presumably, they'd also freeze up. Children are dressed – throughout the winter – in a kind of very thick workman's overall. Scarfs are designed like tunics so that they can be pulled up over the face in extreme cold. And the winter is so predictable that Finns have developed a complex infrastructure to deal with it, unlike in the UK. Almost as soon as it's slippery, lorries appear to grit the roads and pavements. Not long after any heavy snow fall, diggers – whose drivers are, I assume, permanently on call throughout the winter, turn up to deal with the situation. Every year, the sea between Oulu and the island of Hailuoto freezes over, with the ice being so dependably thick that they set up an 'ice road.' It is a 'harsh yet predictable' ecology and you would expect this to select for extreme slow Life History strategists. And, even if it wasn't predictable, it should have selected for intelligence anyway. When it gets down to − 36, you need to be able to make the warmest clothes and the snuggest houses. You must have very long time horizons, planning for the future, being able to conceive of a freezing day during the heights of summer and rationing, and preserving, your resources accordingly. There are so many complex problems that you need to be able to solve if you live in an environment as unforgiving as Finland. And, indeed, you are more likely to be able to solve them if you can cooperate with people. You are more likely to be able to solve them if you are intelligent . . . and that is exactly what Finns are.

2. The Highest IQ in Europe

In many ways, the key problem for the Finns is the nature of their intelligence. It now seems clear that they are the most intelligent native ethnic group in Europe, something which we shouldn't be surprised by considering the conditions to which they are adapted. I led a study on this very issue in 2014 - with Dutch psychologist Jan te Nijenhuis and Oulu-based psychologist Eka Roivainen - which was published in the journal *Intelligence* (Dutton et al., 2014). Every four years, the OECD administers cognitive tests to representative samples of 15 year olds from OECD countries. These are known as PISA – Programme for International Student Assessment – tests, and Finland's consistent high score on them tends to generate a great deal of tiresome and ill-informed media coverage, based on the empirically inaccurate assumption that as ability is surely 100% a matter of environment, so, therefore, everyone should imitate the Finnish education system. The reality of these tests, however, is far more interesting. As we demonstrated in our study, national IQ correlates with national score on PISA at about 0.82, meaning that, to a great extent, PISA – which measures ability in native language, Mathematics and science – is, essentially, an IQ test. Indeed, PISA correlates with other student assessment tests at about 0.9 (Rindermann, 2007). Accordingly, although if you took a PISA test and then took a standard IQ test you would notice that they were very different, PISA is, for all intents and purposes, an IQ test, and we have already observed how IQ is very strongly genetic. We took the Finnish PISA scores for the year 2012 using only native students, both of whose parents had been born in Finland. Using these scores, Finns had the highest IQ in Europe with a score of 102.2 points, which is 2 points higher than the European average.

But PISA wasn't the only measure that revealed high IQ among the Finns. Richard Lynn and Tatu Vanhanen (2012) calculated IQ scores for a large number of countries by combining their mean score in a number of student assessment tests with large representative samples. This gave Finns an IQ score of 105.4, commensurate with scores in Northeast Asian countries. We then compared Finns to Americans (IQ; 100) on the WAIS IV IQ test. Based on this, the Finns had an IQ of 103. Finally, we looked at Finnish average reaction times. As we have already discussed, reaction times are a robust and reliable measure of intelligence and the faster they are, the more intelligent you tend to be. Finns have by far the fastest reaction times of European or European-population countries for which we had data. Converting these into IQ scores, the Finns' lightning reaction times imply an IQ of about 111 points. This is essentially the same as the 112 points attributed to the Ashkenazi Jews, who are the most intelligent ethnic group in the world. Furthermore, Piffer (2019) has examined the percentages of European populations that carry alleles that are associated with very high IQ. Finland, in this respect, is a considerable outlier by European standards, with the Japanese having a higher percentage still. On this basis, Piffer calculates that the Finnish IQ is around 102 and that it is high for genetic reasons. This, in fact, shouldn't be a surprise because it has been shown that, in general, IQ differences between groups are on *g,* and *g* is strongly genetic (Kura et al., 2019).

3. Why is Finnish IQ So High?

So, no matter what the famous modesty of the Finns might tell them to think, they are the most intelligent native ethnic group in Europe. Why is this the case? It may be that it is a relatively recent phenomenon, which would be consistent with the way in which Finland was, until World War II, one of the poorest countries in Europe. Indeed, in 1915, it was officially *the* 'poorest country in Europe' (Shaw, 1915, p.615). I have personally been shocked, talking to Finns born in the 1930s and 1940s, that many of them appear to have suffered the loss of a sibling due to childhood illness. In England, you have to go back to the generation born in around the 1870s to find similar evidence of child mortality. I've looked into my own family history in some detail and none of my parents, grandparents or even great-grandparents witnessed one of their siblings die during childhood. However, this *is* something one finds the generation before that. Finland's poverty in 1915 is partly explained by the fact that it industrialised later than other countries in Western Europe and until as recently as the 1950s it was a predominantly agrarian society (Alestalo & Toivonen, 1977) rather than an industrial one.

As myself and my colleague Michael Woodley of Menie have explored in our book *At Our Wits' End: Why We're Becoming Less Intelligent and What It Means for the Future* (Dutton & Woodley of Menie, 2018), industrialization sets off a process which reduces intelligence. As we have seen, intelligence is strongly genetic and is associated with wealth and high socioeconomic status. Until the Industrial Revolution, it, therefore, predicted the number of surviving offspring you had. The richer half of European populations had about 100% more surviving offspring than the poorer half. As mentioned before, IQ is the strongest predictor of wealth, and that wealth is not arbitrarily distributed. Of course it happens that some children of rich parents are duller than average, and that some children of poor parents are unusually bright. This is however not typical, because IQ is so highly heritable. So, preindustrial conditions meant that intelligence was increasing every generation. However, industrialization saw the price of food and materials collapse as well as massive improvements in medical science, including the widespread use of inoculations as well as increasingly effective treatments for disease. Accordingly, child mortality, which had been about 40% in 1800 though much higher among the poor (and thus among those of low intelligence) nose-dived.

By 1900, wealth no longer predicted higher numbers of surviving offspring. Indeed, the association between intelligence and fertility had now become negative due to several factors – all indirectly rooted in industrialization. Contraception was developed in the late-nineteenth century, and it began to be used by those of higher social status. Intelligence predicts the ability to successfully use contraception: Does a woman who is 'on the pill' take the pill at exactly the same time every day, as she is supposed to, or knock it back with a glass of wine in the evening when she remembers to? Intelligence will predict better understanding how the pill works and thus using it more efficiently. In addition, those with low IQ have, in effect, low impulse control, because their time horizons are shorter. Accordingly, they are less likely to use contraception at all and more likely to impulsively have sex in the moment and think about the consequences later. As a result, people with lower IQ tend to have more offspring or also have them younger, meaning more generations. This negative IQ-fertility nexus in industrial societies is around -0.1 has a number of additional contributing factors, which need not detain us here, including welfare, feminism, and immigration. But the key point is that Finland began industrializing later, and therefore we would expect it to enter 'dysgenic fertility' later. Congruous with this, there existed up until the in 1970s a positive correlation between fertility and the level of education (a strong proxy for intelligence) among Finnish men (Dutton, 2012). The negative intelligence-fertility nexus is weaker among males because females strongly sexually select for status and the select *hypergamously*; for men of *higher* status than themselves (Buss, 1989). This is because, as we will discuss later, pregnancy means they have more to lose from the sexual encounter. If the male will invest in them and their offspring, then both are more likely to survive so females have been selected to be attracted to high status males. This means that as females become more educated (reflecting their higher IQ) there are fewer men that interest them, so they are more likely to have no children at all. By contrast, men are less picky, explaining why the male IQ-fertility nexus is weaker and took longer to become negative.

Another reason for a group developing very high IQ may be a particularly severe selection event. In general, famine and disease select for intelligence, because environmental harshness, more generally, selects for intelligence. You are more likely to survive famine or pestilence if you are wealthy, because wealth will mean that you are healthier, you are better able to obtain food, you have healthier living conditions that protect you from disease, and you will be better able to make and orchestrate plans to escape a disease-ridden area. A famine, or plague, can be regarded, in a pre-industrial context, as Darwinian selection for intelligence playing out over a relatively short period of time. Between 1695 and 1697, Finland suffered 'the Great Famine' in which roughly one third of its population was wiped out due to a combination of famine and the resultant pestilence. Then, between 1866 and 1868, Finland endured 'the Great Hunger Years' during which 20% of the population perished (Haarman, 2016, p.93). You would expect these 'selection events' to have significantly raised national IQ, by killing off the impulsive and the poor and thus the less intelligent. Consistent with this prediction, a few generations after the 'Great Famine' Finland underwent its 'National Awakening,' a kind of intellectual, artistic and cultural Renaissance, something which would very much be predicted to occur if a population's average intelligence suddenly increased. It also experienced a religious revival with the development of so-called 'Awakening Groups' who tended to advocate extreme forms of religiosity. This is not a contradiction, by the way, because it has been shown that though there is a negative correlation between religiousness and doing well in IQ tests, due to the imperfect nature of the test there is actually zero relationship between religiousness and *general intelligence* (Metzen, 2012). This religious revival makes sense insomuch as religiousness is predicted by higher General Factor of Personality (GFP) and socioeconomic status is also predicted by high GFP (Dunkel & Dutton, 2016); so the high GFP Finns, and thus the more religious Finns, would have been more likely to survive the catastrophe at the end of the seventeenth century. In much the same way, it has been argued that the Black Death of the fourteenth century, which killed around one third of the European population, would have selected for intelligence, as evidenced in the fact that around 80% of English serfs succumbed to it (Dodds, 2008). A few generations later, Europe witnessed both the Renaissance, implying that intelligence had increased, and the Reformation, demonstrating that religiousness had done so as well (Dutton & Madison, 2018b).

4. Have Finns Always Been Highly Intelligent?

So, it may be that the high IQ of the Finns is due to these more recent factors. They haven't been able to flourish until relatively recently, it could be averred, because the very harsh nature of their environment has meant they've had to focus mainly on survival. In much the same way, the average IQ of the Inuit has been estimated – at 90 – to be about the same as that of Turks (Lynn, 2015), but, clearly, the Turkish contribution to civilization has been much greater, because they haven't had to focus simply on surviving. However, although all of these recent factors are likely to have elevated Finnish IQ relative to that of other European countries, there is reason to believe that the high intelligence of the Finns has a more ancient origin.

This is implied simply by the distribution of Finnish intelligence. The PISA data which we discussed earlier revealed that though Finns have the highest average intelligence in Europe, they have the smallest standard deviation – the smallest difference between the most intelligent and least intelligent people. This means that they have, relative to other countries of similar IQ, relatively few people who are mentally retarded and relatively few people who are 'geniuses.' 'Genius' is partly defined by outlier high IQ (Dutton & Woodley of Menie, 2018), though also by certain personality factors that we have touched upon. The fact that the standard deviation is smaller than in other European countries implies two interrelated possibilities. Firstly, because the environment is so harsh, Finns are very strongly adapted to it, meaning the gene pool is very small, because only the slightest deviation from the norm will result in a failure to pass on your genes. Outliers will occur by genetic chance but are less likely to manifest the smaller the gene pool is. However, this would, at the very least, imply that Finland's intelligence profile is a function of Darwinian selection and is, thus, pre-industrial in origin.

Secondly, in a very harsh ecology those of outlier high intelligence would provide benefits to their group, such as the innovation of important inventions, but there would also be a definite downside in terms of group cohesion, because they would 'think differently' from the rest. This is highly relevant because under harsh Darwinian conditions it has been shown that groups which are strongly cohesive, in which everyone cooperates with each other, are more likely to survive (Dutton & Woodley of Menie, 2018). People with outlier high IQ tend towards having autistic tendencies, such as hypersensitivity to noise, because autistics are obsessed with systematizing and intelligence involves problem solving and thus systematizing (Karpinski et al., 2018). As people become more intelligent, the positive manifold between the different aspects of intelligence becomes weaker (Woodley, 2011a). This means that at very high levels of IQ, a person's intelligence is rather narrow and, like high functioning autistics, they are brilliant at problem solving but highly deficient when it comes to relatively weakly *g*-loaded abilities, such as empathy and social skill. In an extremely severe ecology, it is quite possible that the benefits of outlier high IQ individuals would be overwhelmed by the problems they would cause in terms of group cohesion. This would mean that a group would be more likely to survive if it had high average intelligence but did not produce outliers. However, this strong environmental adaptation would, once more, imply that there is something more to high Finnish IQ than simply industrializing later than other European countries.

Furthermore, there is evidence that all of the Finno-Ugric peoples who are genetic cousins of the Finns have relatively high IQ. Based on intelligence tests ad a series of proxy measures, it has been found that the Finno-Ugric peoples of Russia – despite often living in relatively poor conditions – have higher average intelligence than the Russians, and, in general, the higher the Finno-Ugric share is in a region of Russia or Estonia then the higher is the average IQ of that region (Dutton et al., 2018). This would be congruous, once more, with Finnish high IQ being, in part, a reflection of Finland's cold yet stable environment selecting for high intelligence under conditions of Darwinian selection.

5. High Average Intelligence and the Islamic Invasion

The high IQ of the Finns has many advantages, as we have already noted. It means that Finland is a very safe country and you are most unlikely to be a victim of crime. The other evening, I cycled into the centre of Oulu to go to the pub with a Romanian friend. I left my bike outside this pub, known as Hemingway's, and didn't bother locking it. I never lock my bike, so confident am I that nobody is going to steal it. And Finns are *not* going to steal it because, as a function of intelligence, they are low in criminality. They look into the future, as high IQ predicts they would, and they think, 'If everyone went around stealing, we'd end up living in a terrible society.' Intelligence predicts the ability, albeit weakly, to more easily imagine what it is like to be somebody else, and thus empathise with them. Finns, being cleverer than English people, are more likely to think, 'If somebody stole my bike, I wouldn't like it, so I won't steal their bike.' Criminality is a function of low IQ for this reason and also because criminals will tend to underestimate the chances of getting caught, often because their low IQ makes them impulsive and also because low IQ predicts high self-esteem. Accordingly, they will be over-confident in their ability to get away with stealing a bicycle and they will fail to compute all of the possible means through which they might get caught. Furthermore, in a society with high IQ things are simply less likely to go wrong. There will always be somebody able to solve some sudden, unexpected problem and there will be relatively low levels of social conflict because, high in empathy and trust, people will simply be better able to get along. This makes Oulu, and Finland more broadly, a safe, if dull, place to live.

However, when living conditions are good, and levels of stress and thus religiosity decrease, then the high intelligence of the Finns can be weaponized against them and this is exactly what is happening. Religiousness is strongly predicted by stress. When people are subject to stress then an evolved cognitive bias to believe in god, or something similar, is elevated (Norenzayan & Shariff, 2008). We know that religiousness is an evolved cognitive bias. It has been selected for in prehistory, as evidenced by the fact that it is 40% genetic, it is (seemingly genetically) correlated with health, and it is associated with elevated fertility. It has likely been selected for precisely because it reduces stress and also because, as we have discussed, those who believed they were being watched over by a moral god would be less likely to get into fights and thus less likely to be cast out by the band (Dutton & Madison, 2018b). Religiousness is also associated **with** ethnocentric behaviour; with being prepared to make sacrifices for your group and with strongly distrusting members of other groups (Dutton et al., 2016a). Many computer models have shown that the group that is highest in positive and negative ethnocentrism is most likely to dominate other groups, all else controlled for (e.g. Hammond & Axelrod, 2006). The religions that have survived the process of evolution – or, rather, the religious groups that have done so – are those which take evolutionary imperatives, such as high positive and negative ethnocentrism, and turn these into the will of God. These groups believe that they are a special people, chosen by God, while other groups are in league with the Devil (Sela et al., 2015). In other words, even if a people are highly intelligent, if they are religious then they will still likely be strongly ethnocentric and, thus, more likely to repel the invader.

The problem, however, is if a society's living standards become too high, and its stress levels become too low, then it will gradually stop believing in God and so stop perceiving its own culture as uniquely important and perceiving foreigners as a dire and cosmic threat. Much of Eastern Europe is not religious, but it is poorer than Western Europe, and poverty also predicts ethnocentrism (Dutton et al., 2016). It industrialized later than much of Western Europe which, as will discuss below, it is likely to mean a lower percentage of people with maladaptive ideas, Moreover, their many years of Communism are likely to have left them distrustful of Leftist ideas and so more open to indoctrination with nationalism. Japan has very high living standards and it is not especially religious (Lynn & Vanhanen, 2012, Ch. 10). But it has the further protection, which all Northeast Asians have, of being very high in ethnocentrism compared to Europeans, based on the World Values Survey (Dutton et al., 2016). Further, Neuliep et al. (2001) conducted a survey with American and Japanese students and found that on all measures the Japanese were significantly more positively ethnocentric than the Americans. A study of 79 pupils in Toronto aged 12 to 14 found that East Asian children were more positively ethnocentric than those of other races (Smith & Schneidner, 2000). This ethnocentrism is seemingly genetic. Pursuing this line of research, Cheon et al. (2014) reported that the serotonin transporter gene polymorphism (5-HTTLPR) is associated with ethnocentric behaviour. Further research has found that 70-80% of an East Asian sample carried the short form of this gene; that is to say the form that makes you *more* ethnocentric. Only 40-45% of Europeans in the sample carried the short form of the gene. Indeed, it was found that across 29 nations, the more collectivist a culture was the more likely it was to have the short form as the prevalent allele in the population (Chiao & Blizinsky, 2009).

So, Eastern Europeans are currently protected by their relative poverty and perhaps also the recent trauma of having been subject to invasion and occupation. Northeast Asians have a further protection mechanism when religiousness collapses. Western Europeans, including Finns, lack these protections. As such, when stress drops too low, this will set off a spiral of religious destruction, because it will become more acceptable to question and ridicule the religion, meaning it will lose its sense of awe and its capacity to inspire. The religion will cease to be uniquely real and the culture it protects will cease to be uniquely significant, with that culture's unique significance not protected by the genetic ethnocentrism you find in countries such as Japan. High living standards will also mean that the level of purifying natural selection pressure against poor genetic health is extremely low. This will mean that, every generation, a higher and higher proportion of the population will carry mutant genes which would have been selected out under pre-industrial conditions because they would have resulted in a suboptimal immune system and thus dying, as a child, of measles or typhus or some other now curable disease. Before the Industrial Revolution, before effective medicine, child mortality was 40%. It is now approximately 1%. Mutations of the body are comorbid with mutations of the mind as the brain constitutes 84% of the genome it is acutely sensitive to mutation. Thus, Michael Woodley of Menie and his colleagues (Woodley of Menie et al., 2017) have proposed what they call the 'Social Epistasis Amplification Model.' They argue that under conditions of weakened selection, there would be more and more per capita individuals with higher and higher mutational load, resulting not just in bodies but in minds which would have been maladaptive under Darwinian conditions due to suffering from assorted neurological disorders such as autism and schizophrenia. Some of these people will have 'spiteful mutations' which will make them, for example, want to not breed, or be confused about their gender, or want to destroy their own ethnic group. But this will cause problems far beyond their own capacity to pass on their genes.

Woodley of Menie (*The Jolly Heretic,* 11th February 2019) has observed that human society might be more usefully compared to a beehive than to a group of highly advanced chimpanzees. This comparison works because, like bees, humans live in intensely social societies in which there are extremely clearly marked divisions of labour. In addition, human society – like a beehive – is in stark competition with other similar societies, meaning it is subject to a very high level of 'group selection.' If there is competition between individuals, then you have 'individual selection.' However, you can pass on your genes indirectly, if there is competition between ethnic groups (which are genetic extended families), by investing in that group. This what is meant by group selection.

6. The 'False Allure of Group Selection'

Group selection has also been criticized in depth by the popular psychologist Steven Pinker (18th June 2012) in an essay entitled 'The false allure of group selection.' Dealing with these criticisms requires a rather technical aside, but, alas, if we don't deal with them then there will be those who will be unconvinced by everything that follows, so the technical aside is required.

Dutton et al. (2018) observed that Pinker's 'key criticisms are that (1) Group selection deviates from the "random mutation" model inherent in evolution; (2) We are clearly not going to be selected to damage our individual interests, as group selection implies; and (3) Human altruism is self-interested and does not involve the kind of self-sacrifice engaged in by sterile bees.' Each of these points, note Dutton et al. (2018), can be answered. 'Firstly, if the group selection model is building on the individual selection model then it is bound to present a slightly different metaphor. To dismiss it on these grounds seems to betoken a fervent attachment to the original idea' (Dutton et al., 2018), a kind of fallacious appeal to authority or to the *status quo*. Moreover, in what way does group selection deviate from this 'model'? It can be countered that it does not. Non-random association in forming a group is important to create differences between group averages and is therefore expected to enhance group selection.

Secondly, proponents of group selection argue that the group selection model merely suggests that a group will be more successful if an optimum percentage of its members are inclined to sacrifice themselves for their group. So, it is argued that a group will have higher fitness if there is sufficient genetic diversity, such that an optimum percentage are inclined to damage their individual interests for the sake of the group. In much the same way, a fertile mother who sacrifices her life to save her children is sacrificing her individual interests for those of her kin group; the model is merely extending this to a much larger kin group, the 'ethny'. Thirdly, 'it is clearly the case that a small percentage, in many groups, is indeed prepared to sacrifice itself for the group' (Dutton et al., 2018), as has been explored in depth elsewhere (e.g. Salter, 2007). It is also worth noting that Pinker entitles his critique of group selection, 'The false allure of group selection,' as if those who regard the concept as reasonable have been somehow beguiled and bewitched, thus 'poisoning the well' of academic debate. In effect, the reader will be nudged to believe that those who accept group selection are deceived. In conclusion, the concept of group selection is perfectly acceptable and if you are part of a tightly structured, cooperative group in competition with other groups then this process is taking place.

7. Of Bees and Men

Woodley of Menie (*The Jolly Heretic,* 11th February 2019) argues that cooperation is so central to human, and bee, society that, in many crucial respects, these societies can be conceived of as kinds of organisms, with each individual playing a small but important role in the optimum functioning – and thus survival – of that organism. Every individual is part of a profoundly interconnected network wherein he relies on those with whom he interacts to behave in an adaptive fashion such that his own genes are expressed, phenotypically, in the same optimally adaptive fashion. It follows that if a feeder bee – in charge of feeding the larvae – has a mutation which causes her to feed the larvae at random, rather than according to an instinctively 'normal' pattern, then she will damage the fitness of the entire hive, because there will be, for example, far too many queens and not enough workers. Similarly, if a mutant human carries a spiteful mutation which makes him believe, and propagate the view that, life has no meaning, then his mutation will impact those who associate with him by causing – as with the bees – their environment to be different from that which their genes are optimally adapted to. This will mean that their genes will be expressed differently; sub-optimally. These non-carriers of spiteful mutations may be somewhat genetically predisposed to be Nihilists and abandon reproduction 'because life's pointless' due to certain environmental conditions. If these conditions never occur, because the spiteful mutation carriers are not present in their society, they will not develop those destructive behaviours. But once the spiteful mutant manifests, then he can undermine the societal culture – such as its religious rituals – and stop his genetically 'normal' co-ethnics from being religious, meaning that his maladaptive worldview can spread like a plague even to those who lack the spiteful mutation. And, so, this single mutant – even more so if there are many of them

– can help to undermine the extent to which the group is optimally group-selected.

If these spiteful mutants ascend to positions of power, they will be able to influence those who do not carry the spiteful mutations - by persuading them, for example, that women who are dedicated wives and mothers are losers – to a very significant degree indeed, by heavily undermining the capacity of the group to engage in adaptive behaviour and hold adaptive beliefs. In that religiousness was selected for under preindustrial conditions, we would, therefore, expect atheism to be correlated with evidence of mutation. We have shown that this is indeed the case in our study 'The Mutant Says In His Heart, "There is No God"' (Dutton et al., 2018b) that this is indeed the case. It is predicted by autism. It is predicted by left-handedness; this is a marker of developmental instability and thus mutational load, because we are evolved to be symmetrical and symmetrical brains tend to result in right-handedness. Atheism is associated with poor physical health, poor (genetic) mental health and even physical asymmetry; that is not being physically symmetrical. Political conservativism crosses over with fundamentalism at about 0.75 (Laythe et al., 2001) and political conservatives in the USA have more symmetrical faces than political liberals (Berggren et al., 2017). The essence of beauty is symmetry because, as already noted, we are supposed to be symmetrical, so if we are not then this indicates mutational load either directly - mutations have made us less symmetrical - or indirectly - mutations have compromised our immune system so we cannot maintain a symmetrical phenotype because we are compelled to use proportionately more of our bioenergetic resources to fight off disease, especially in childhood. In positions of power, these mutants can then persuade even non-carriers to reject religiosity and even adopt some ideology that destroys their ethnic group, and thus their own genetic interests, such as Multiculturalism.

And it gets worse because, under Industrial conditions, we would actually expect these 'spiteful mutants' to ascend to the top of society. Under preindustrial conditions, there was extremely intense selection for wealth. As we have discussed, in England, for example, in the seventeenth century, the completed fertility (number of surviving children) of the richer 50% of the population was double that of the completed fertility of the poorer 50% of the population. Due to the robust connection between intelligence and wealth, this elevated the intelligence of the population every generation as those at the bottom died out and there was social descent to fill the positions they vacated. For reasons we have discussed, such as contraception, this association has now been reversed. However, socioeconomic status is about 70% genetic across generations in societies as different as India and Sweden (Clark, 2014). Accordingly, though people do move up and down the social hierarchy, there is a significant degree to which socioeconomic status is genetic. The key predictors of socioeconomic status are intelligence and personality, specifically being high in Conscientiousness or, more generally, being high in the General Factor of Personality. However, various optimum combinations of these traits will also predict high social status. Most obviously, a person could be relatively low in GFP but more than compensate for this via stratospherically high intelligence. However, mutational load only has a very small negative impact on IQ (Woodley, 2011b).

As such, it can be argued that the highly intelligent – and thus wealthier – people who compose society's elite will have been under weaker purifying selection pressure against mutant genes for longer. Living in better conditions, their child mortality levels will have collapsed before those of the caste at the bottom of society, meaning that members of this elite caste will have more mutations and, specifically, more spiteful mutations. With the traditional elite of the Victorian era failing to breed in significant numbers, those who were from the middle of that society have moved up and their descendants occupy elite positions, while a very low IQ underclass, which didn't exist in the Victorian period, is now at the bottom, the late Victorians having had an average IQ about 15 points higher than ours (Dutton & Woodley of Menie, 2018). This helps, in part, to explain why maladaptive ideas tend to emanate from the elite while 'traditional' (i.e. 'adaptive') ideas – such as nationalism – are still held to by the working class. 'Traditional' ideas would tend to be those which went relatively unquestioned prior to the Industrial Revolution: belief in God, the maintenance of custom as good in itself, the acceptance of societal hierarchy, reverence for the ancestors and their achievements, and loyalty to family, region and nation (see Scruton, 2002). Holding adaptive ideas in the working class would be further elevated by our cognitive bias towards such ideas under stress, with those of low SES tending to live more stressful lives. One of the maladaptive elite ideas is to reject traditional, hierarchical religion, a spiteful mutation the consequences of which they spread throughout the society. As if this wasn't problematic enough, the more complex the society is the more sensitive it would be to mutation because – to return to the organism metaphor – the more interactive and reliant upon each other's optimum functioning the different components of the organism would be.

8. Of Mice and Men

Woodley of Menie et al. (2017) have highlighted John B. Calhoun's (1917-1995) famous 'Mouse Utopia' experiment at the University of Maryland between 1968 and 1973 (Calhoun, 1973) as stark evidence of this process of Social Epistasis Amplification. The mice were placed in conditions where there was, essentially, no purifying Darwinian Selection: no inclement weather, no predation, no serious illnesses, and ample access to nutritious food as well as constant medical care. This paralleled the way in which the Industrial Revolution brought inoculations, effective medicine, cheaper food, and an ever-improving standard of living. As with humans, there was an initial population explosion, but then growth began to slow down as fewer and fewer mice had children. Eventually, the population plateaued and, as with developed countries where there is little or no immigration such as Japan, it then went into decline. This was paralleled by the observation of very unusual forms of behaviour. Females expelled their young from the nest too early so that they weren't sufficiently socialised to be able to deal with other mice. Females also became increasingly masculinised, aggressive and uninterested in breeding, even attacking their own young. Slightly later, a startling group was noticed among the males. Known as 'the Beautiful Ones,' they were remarkably effeminate. They didn't fight for territory, had no interest in females, and spent all of their time grooming each other. Calhoun (1973, p.86) described them thus:

> 'Autistic-like creatures, capable only of the most simple [*sic.*] behaviours compatible with physiological survival, emerge out of this process. Their spirit has died ('the first death'). They are no longer capable of executing the more complex behaviours compatible with species survival. The species in such settings die.'

58

Eventually all the mice were either 'beautiful ones,' masculinized females or socially inept. Consequently, no more mice were born, and the colony died out.

Woodley of Menie et al. (2017) show that the colony was nowhere close to being overcrowded when the population growth began to slow down. Thus, the simplest explanation is that mutant genes, causing maladaptive behaviour in mice, were no longer being expunged from the population. This led to mutational meltdown, where the percentage of maladapted 'mutants' in a population is so high that the society collapses. It is, of course, striking that one of the issues of our time is deviant sexuality and transsexuality in males and females, something that is associated with significantly genetically influenced physical and mental illness and thus with high mutational load (Blanchard, 2008). A team of Woodley of Menie's (Bachmann et al., 2018) have since worked out, in more detail, why the colony collapsed by recreating a mouse colony and then introducing mice whose overexposure to a particular pheromone causes them to replicate much of the behaviour patterns of the Beautiful Ones. Their presence caused the normal mice to behave in the manner in which Calhoun's mice had behaved when the Beautiful Ones had been present. The Beautiful Ones were causing the normal mice's genes to be sub-optimally expressed, leading to maladaptive behaviour, and, indirectly, to an inability, or lack of desire, to breed. By the end of the experiment, females would have to pester males for sex, and the males wouldn't be remotely interested. As with humans and social epistasis amplification, the mutant mice compromised the delicate relationship between the different mice in the mouse super-organism wherein one mouse's genes could only be optimally phenotypically expressed if that was true of all the other mice in his or her network did so. The Beautiful Ones, Woodley of Menie's team found, were producing a pheromone which made all the other mice feel extremely anxious – as might occur if humans were convinced there was no God – by virtue of it overwhelming the scent of normal urine. This meant that normal mice could not read the signals in that urine of whether friend or foe had marked the territory as theirs. Consequently, this meant that they could not establish social hierarchies or coalitions and that they could not work out who they could safely interact with, and the anxiety this induced reduced their testosterone levels. These 'mutants' thus caused the colony's collapse. However, when the mutants were removed from the mouse community then everything very quickly returned to normal.

9. Which Strategy? Ethnocentrism or Genius?

So, returning to humans, when all this happens – when religiousness collapses - then a society that has high average intelligence is in very serious trouble. Of course, there will be variations in how ethnic groups have evolved, meaning there will be variation in how much trouble they are in. Japan, for example, is an extremely wealthy society and is no longer particularly religious (Lynn & Vanhanen, 2012, Ch. 10). Nevertheless, it remains much more ethnocentric than most European nations. This is likely because the extreme harshness of its ecology very strongly selected for extraordinarily cooperative social groups who were prepared to act lethally against outsiders and also because the severity of its ecology rendered geniuses an intolerable liability, due to their tendency to be anti-social dreamers (Dutton & Charlton, 2015). Thus, as I have argued in my book *Race Differences in Ethnocentrism* (Dutton, In Press), European nations had, at least, two evolutionary strategies open to them as means of maximising their genetic fitness: (1) The highly *Ethnocentric Strategy*, and (2) The *Genius Strategy*. The latter involved being low in negative ethnocentrism. This allowed the group to expand, increase its gene pool, and so throw up geniuses by genetic chance. These geniuses would then innovate things of great use to their group, allowing the group to expand further. This would be adaptive as long as an optimum level of ethnocentrism was maintained by means of religion. This option was simply not open to the Japanese because the environment was too harsh for the gene pool to diversify to the extent required to produce a significant per capita number of geniuses and the risk factors involved in geniuses – such as low IQ anti-social people manifesting by chance and destroying group harmony – were simply too high.

Thus, the collapse of religion is problematic for all intelligent societies, but especially for European ones, because they have followed this 'genius strategy.' When religiousness declines, God - who for so long has protected the society from the negative consequences of having high intelligence – is dead. Accordingly, the society's intelligence leads to its own destruction. Most importantly, intelligence predicts being trusting. There is a naivety to being intelligent. High in trust, highly intelligent people will let foreigners into the society and assume that they are ultimately honest and good and everything will be okay in the end. Low in self-esteem, they will perceive their own culture as worthless and backward and not worth preserving. They are likely to believe that their culture will be improved by being 'enriched' by somehow superior elements of the cultures whose adherents are entering their country. Registering high in Openness, those of high intelligence will be excited and fascinated by the immigrants and they will be prone to try to look for the positive in them. Being high in empathy, if the immigrants, or potential immigrants, are from poorer societies then the intelligent will be strongly inclined to help them. Being trusting, they will assume that this will be reciprocated; that the invaders will be extremely grateful. Denuded of their religion, the highly intelligent are woefully naïve. They are decadent and they are invaded by the Enemy at the Gate.

And this is Finland's problem. It is too intelligent and, therefore, once stripped of its nationalism-inspiring religion it is, as a nation, that naïve 26-year-old girl walking home through Otto Karhi Park in the centre of Oulu on that night in December 2007. It has less to protect itself with than most other countries. And the consequences are painfully obvious. But as if Finland's high IQ wasn't a big enough problem in protecting itself from some of those of 'foreign background,' it's modal personality makes the situation even worse. However, as we will now explore, it is quite clear that Finland operates a far less pronounced 'Genius Strategy' than other European countries. Indeed, in many key respects it is an outlier and appears to be more similar to an ethnocentric society such as Japan. The Japanese are well known for being cooperative and altruistic but also for being prepared to make incredible sacrifices for their group and, similarly, for being ferocious towards out-group members who seriously threaten them. Their treatment of British soldiers in Southeast Asia during World War II is a testimony to this (see MacArthur, 2006). We will note that History suggests that the Finns follow the same evolutionary strategy, though to a much less pronounced degree.

Chapter Five

The Nicest People in Europe and the Least Creative

1. The Average Finnish Personality

American psychologist Lee Jussim (2012), whom we met earlier, has shown that stereotypes have a very high degree of accuracy. Thus, if there is a commonly agreed stereotype about the average Finnish personality it is likely to be empirically accurate. Scientific purists can dismiss it as 'impressionistic' or as a fallacious 'appeal to anecdote' if it makes them feel better, but few people are going to be convinced by this. I've spoken to many expatriates over the years, living in Oulu. At one point, I was tasked with interviewing a different expatriate entrepreneur every week for a series of articles for an English language online newspaper run by Oulu city council. I always asked them about what they thought of Finns and I would always almost be met with the same kinds of words: 'shy,' 'honest,' 'meticulous' and 'hard working.' I also found that most expatriates that had interacted with Finland's 5.4% Swedish-speaking minority (hereafter Finland-Swedes) found them to be much more 'confident' and 'outgoing' than the Finns.

In my own anthropological fieldwork with Finns, published in a book called *The Finnuit* (Dutton, 2009b), I conceptualised Finland as an Arctic, rather than Scandinavian, culture. When speaking to my informants about the nature of Finnishness, the same kind of terms appeared again and again, and they starkly paralleled how foreigners seem to view Finns, as well as how Arctic groups, such as the Greenlandic, are perceived: 'low self-esteem,' 'melancholy,' 'shy,' 'honest,' and 'we don't speak much.' Finns also concurred with what expatriates whom I had interviewed thought about Finland-Swedes. These appear to be strongly robust stereotypes. A travel writer reported, in 1889, that Finland-Swedes were more 'lively' and versatile' but less 'meditative' and 'profound' than Finns (Reuter, 1889, pp.61-66). Finnish students, interviewed in 1967, claimed that Finland-Swedes were 'active' while Finns were emotionally 'strong' (Kivisto & Makelä, 1967). This is congruous with the Finnish term *sisu* ('guts'), an aspect of which is not just bravery but remaining stolid and strong in the face of adversity; a kind of Stoicism (Dutton, 2009b).

62

These comments can be reduced down to the Big 5 personality traits which we discussed earlier. They imply that Finns – relative to, for example, the British – are low in Extraversion, high in Conscientiousness, high in Agreeableness, and high in (at least aspects of) Neuroticism. In a study published in 2016 (Dutton et al., 2016b), we tested this by comparing a large sample of Finns to a large sample of Finland-Swedes on the results of a personality test. This didn't suffer from the problems usually inherent in making intercultural comparisons on personality tests because Finland-Swedes, being such a small minority, are strongly integrated into Finnish culture, all those tested spoke Finnish, and, thus, they will not simply be comparing themselves to other Finland-Swedes, but instead to Finnish citizens more broadly. We found, congruous with the stereotypes, that the Finland-Swedes were higher in Extraversion and lower in Neuroticism. The Finland-Swedes were also higher in Conscientiousness than the Finns and scored higher, overall, on General Factor of Personality. They did, however, have slightly lower average IQ than the Finns, implying that their very significant over-representation among the Finnish elite is a reflection, in part, of their high General Factor of Personality. Congruous with this evidence of low Finnish Extraversion, Tulviste et al. (2003) compared samples from three supposedly 'quiet' (and so introvert) nations – the Swedes, the Estonians and the Finns – and found that Finns are by far the least talkative and the most reserved, even when talking to their own babies. There is, by the way, little question that Finland-Swedes are overrepresented among the Finnish elite. As noted in Dutton and colleagues (2016), 'in 2011, 24% of board members of the 50 largest Finnish companies were Finland-Swedes.'

In terms of direct personality research, there is one other important finding, the full significance of which will become relevant shortly. We have already noted that the IQ range among Finns – the difference between the least and most intelligent Finn – is smaller than in other European countries. Fascinatingly, their personality range is also narrow. There is a very low percentage of Finns, relative to the other 54 countries examined in a study to which we will now turn, who are outliers in terms of their personality: very high or low in Extraversion compared to the general population, for example. Lukaszewkski et al. (2017) have presented the Socioecological Complexity Hypothesis. According to this model, personality trait covariance is shaped within a particular social environment and the more technologically advanced an ethnic group is then the narrower is the degree of each member's level of specialization, because as societies become more technologically complex we move from everyone hunting to people doing very specific jobs. They hypothesize that the more specialized the society is then the more niches there are to fill. The more niches there are to fill then the greater will be the number of specific personality types that are selected for. Thus, one niche might benefit from a very specific combination of relatively low Agreeableness, relatively low Conscientiousness, and relatively high Neuroticism, as has been proposed to be the case with many artists (Post, 1994). In a less advanced society, there are fewer niches, meaning less intense selection for highly specific personality types, leading to more positive correlations between more aspects of personality. Lukaszewksi et al. have found, based on analysis of data from Finland and 54 other countries, that aspects of personality are more strongly inter-correlated the less socioeconomically advanced the nation is, as measured by economic development and urbanization. Such a model would, therefore, predict that there would be considerable personality diversity in Finland, and, indeed, the authors record Finland as being one of the most technologically advanced countries in the world. Nevertheless, Finland is a pronounced negative outlier in terms of the theory. Different aspects of personality are much more strongly correlated in Finland than in almost all other Western or European countries. Finland exhibits a very narrow range of personality; the level of personality diversity is similar to that found in third world countries such as Mexico or Ethiopia. This means that there is a very limited range of personalities in Finland. In personality terms, Finns are all very similar to each other when compared to other Western countries, even including Japan.

2. SchizoFINNia

I never really understood what schizophrenia was until I travelled to Finland. That Finnish man that I tried to engage in conversation on the plane in July 2003 may well have been 'shy.' But it is quite possible that his shyness was a symptom of schizoid personality. He sat there, reading some novel or other, divorced from reality for the entire flight. He was completely indifferent to my overtures of friendliness – no reciprocal smile, hardly any eye contact, no human warmth. This man was completely indifferent to the fact – which he surely should have known if he'd spent time in England – that he was just being plain rude and standoffish. Emotionally cold, joyless, and yearning to just be left alone, he manifested most of the symptoms of schizoid personality: an extremely mild form of schizophrenia.

After I moved to Finland, I began to notice just how common this kind of personality was, especially among men. To meet people like this was so unusual in England that it would be remarkable. But in Finland, it was me who, as a relatively outgoing and talkative man, was worthy of comment. Most men I met in Finland appeared to be these silent, unemotional types. Their symptoms were not as pronounced as those of the man on the plane, perhaps, but they were the type of men of whom I met very few in England: Taciturn, introvert, joyless, reserved, and perfectly happy to be solitary, engaged in pursuits that were absolute anathema to me such as hunting, trekking and cross-country skiing. Presumably their incapacity to experience joy rendered sport quite attractive, because it would elevate their endorphin levels. It also appeared to render alcohol extremely inviting.

I don't think I really understood what alcoholism was until I came to Finland either. In England, it was something you'd joke about. If somebody at university drank very heavily then you would playfully mock them as being an 'alcoholic.' And even adults I knew who were 'boozers' (the euphemism for an educated alcoholic) would have no problem holding down a job or avoiding interaction with the police for drink driving. They'd simply be jolly red-faced Chartered Accountants who'd respectably die of cancer in their late-fifties. But in Finland the situation was much more extreme. Pubs, at least in less well-off parts of Oulu, would be patronized by these 'silent' men who would be allowed, by the landlord, to drink quite spectacular quantities of alcohol, at which point they would be sufficiently pissed to experience 'joy' for a while, before being bundled into a taxi. One night, in July 2005, I had to clamber over a bald middle-aged man who was lying in the street, unconscious, in front of a bar just outside the centre of Oulu. In February 2009, the chef we had employed to do the catering for my daughter's christening – the silent, joyless type – turned up late and completely smashed, incapable of doing anything. Fortunately, my sister-in-law's common law husband, himself a chef, was able to rescue the situation. I heard that the alcoholic chef died shortly afterwards, aged 44. Alcoholism may be prevalent in Finland because Finns took up complex agriculture relatively late, concentrating on 'slash and burn' cultivation. As a consequence, Finns would have begun to manufacture alcohol to a serious extent relatively late, as the discovery of alcohol is associated with the cultivation of fruit. Once this discovery is made, there is an advantage to being adapted to being able to cope with alcohol because it means you don't have to drink the water, which will often be dirty and full of pathogens. So, genes allowing you to not become drunk, not become depressed as a consequence of drink, not get hung over, and not become an alcoholic will spread. The closer you get to the Fertile Crescent in the Middle East, where agriculture began 10,000 years ago, the longer they have had to spread, which is probably why young Greek men seem to spend their Friday evenings strutting around in very fashionable clothes rather than getting utterly hammered. In Finland, these genes have had less time to spread and in somewhere like Greenland there is no adaptation to alcohol whatsoever, because the Inuit never developed farming (Cochran & Harpending, 2009). Accordingly, there are very high levels of alcoholism among the Greenlandic (Dutton, 2009, p.107). In addition, East Asians are poorly adapted to alcohol because they learnt to purify water using tea as well as alcohol. Many studies have found that Finns are high on specific East Asian genetic markers compared to other Europeans. This has been found on analysis of Finnish Y-chromosomes (Zerjal et al., 1997) as well as in examinations of Finnish mitochondrial DNA (Meinilä et al., 2003) and in genomes extracted from Finnish skeletons buried between 200 and 3,500 years ago (Lamnidis et al., 2018). Estimates of Finnish East Asian admixture vary. One (outlier) study puts it at 30%, Wiik, 2006), others in the in the range of 10% (Gugliemino, 1990) while others still suggest just 2% (Norio, 2003, p.261). However, any such Asian admixture may be germane to making sense of the noticeably high levels of problems with alcohol in Finland. There is a so-called J-curve from Saami-land (Northern Lapland) downwards through European societies with substantial problems with alcohol and suicide. This follows the path of Finn-Ugric genetic influence (Marusic, 2005).

But returning to schizophrenia, it is clear from surveys that it is no coincidence that I should have noticed so many of these schizoid males in Finland, drinking themselves in stupors in bars. As we have already discussed, schizophrenia sits at one end of a spectrum, low empathy is positioned at the other, and 'normal' is somewhere in the middle. Schizophrenics over-empathise. Schizophrenia is associated with being very bad at theorizing, as suffering from it involves 'jumping to conclusions,' and 'bias against disconfirmatory evidence' (Crespi, 2016). Finland has a high rate of schizophrenia, at least relative to the UK. A twin study of schizophrenia in Finland found the disorder to be 83% genetic (Ekelund et al., 2000). In a representative sample of 47,445 Finns born 1972–1984, it was discovered that 0.7% of Finland-Swedes but 1.5% of Finns had been diagnosed with a schizophrenia spectrum disorder (Suvisaari et al., 2014). This is particularly telling as both populations are well integrated into the same national state, and are exposed to the same environment and health care system. The percentage of the population with schizophrenia in the UK, Greece or Iceland is approximately 0.5% while in Sweden it is about 1%. Part of the cause of schizophrenia is environmental stress, meaning we would expect it to be high in poor countries, and it is (Read, 2010). But Finland has a high rate by the standards of developed nations, although it is massively exceeded by wealthy Northeast Asian countries, where the prevalence is about 4.5% (Dutton et al., In Press).

These differences in schizophrenia prevalence might not seem large, but small differences at the extreme of a spectrum will ultimately alter the average, meaning that they can have quite substantial effects. If 2% of a country's population have an IQ of above 130 points, 130 being the average IQ of a Physics PhD researcher in the 1960s (Harmon, 1961), then the average IQ of the population is 100; the average IQ of Britain and, indeed, most of Western Europe. If we change this, so that less than 0.1% of the population have an IQ of over 130, then, based on a normal distribution, the average IQ drops to 85, this being the mean IQ score in many Middle Eastern countries and also the average IQ of African Americans (Lynn, 2015). It has been comprehensively shown that national IQ robustly predicts every aspect of civilization: adult literacy at 0.6, research and development levels: 0.6, GDP: 0.65, severe poverty levels: -0.7, corruption: -0.68, democracy: 0.53, women's rights: 0.3, poor health: -0.5, life expectancy: 0.7, infant mortality: -0.7, murder rate: -0.82 and so on (Lynn & Vanhanen, 2012). Just as with IQ, small differences in the schizophrenia rate are important because they reflect a country's different average position on this spectrum. This placing, as with a country's IQ placement, will, therefore, lead to one noticing considerable differences in everyday life when comparing the two societies.

3. My Personal Experience of a High Schizophrenia Society

The average Finn can be said to be higher on the schizotypy spectrum than the average English person. The average Finn will, thus, be more likely to display symptoms of schizoid personality than will the average English person. The result is the stereotype that Finns are taciturn, emotionless, joyless, and superficially cold. However, there are other subtler dimensions that you would also expect to observe. Finns would be lower, on average, in systematizing than would people in countries with lower average levels of schizophrenia. As we have discussed, this would make them prone to jumping to conclusions, to being biased against 'disconfirmatory evidence' and, in the face of signals potentially indicating something negative, to over-analyze and to become paranoid. I had my own experience of this between 2016 and 2017. In 2013, I had written an academic article (Dutton & Lynn, 2013) on Finland's declining IQ – declining national IQ is known as the negative Flynn Effect or negative Lynn-Flynn Effect after those who discovered it (Kanazawa, 2012, p.188). The paper was based around a secondary analysis of a table in a Finnish-language Master's thesis which had been sent to me by a researcher at the Finnish Military Academy in response to my request for army IQ data. This researcher informed me that the information I needed was on this particular page of the thesis. In that the thesis was in Finnish, my Finnish is only conversational, and I'd been told where to look, I assumed this was simply the military's data which the author was presenting and discussing in some way, but I cited the thesis in the paper nevertheless just to be absolutely clear. Prof. Richard Lynn, my then boss and coauthor on the paper responsible for all the statistical work, removed the citation, explaining that as it was 'the military's data' we didn't need to cite it. I'd never written a quantitative paper before at that stage, so I assumed he knew what he was talking about, especially as he was experienced in this area of research, was far more senior than me, and was even on the editorial board of the journal, *Intelligence,* where it would be published. This meant that we had cited the thesis for some points – written in Finnish, though clear to me – but not for the table containing the army's IQ data. The article was duly peer-reviewed and published in *Intelligence,* a leading journal, and even reported on in the Finnish national press. I also did another paper with Lynn, in 2014, using military data, sent to me by a particular researcher, from within his Finnish language paper (Dutton & Lynn, 2014). Lynn analyzed this and returned it with some numbers which I had no reason to doubt, having no training in statistics.

Then in summer 2016, an article appeared in *Helsingin Sanomat* accusing me of having somehow stolen the military's confidential IQ data. The article made clear that I was a race realist as was Lynn and that Lynn had worked with the race realist Tatu Vanhanen who is a bogeyman for the Finnish Left and, thus, for much of Finland's media and academia. There was a full – surely utterly paranoid and pointless - investigation into this, despite my explaining to them what I have just explained. This investigation, naturally, concluded that there had been no leak. The author of the Master's thesis and the academic who'd sent me his Finnish-language paper then made a formal complaint to Oulu University accusing me of plagiarism. They were able to do this because in 2011 I had been made 'Docent of the Anthropology of Religion and Finnish Culture' at Oulu University. A 'Docent' is a bit like a 'Reader' in the UK, except it is a qualification rather than a job. It makes you a 'Senior Researcher,' based on an evaluation of your published research, but it also renders you affiliated to the department. Armed with it, you can go on to apply to be a 'Professor,' except that a 'Professor' is different in Finland because he is always a head of department. In the UK, someone can simply be made a 'Professor' based on their research quality. So, the Finns' 'official translation' of 'Docent' is 'Adjunct Professor' while the Swedes', probably more accurate, translation is 'Reader.' As part of their 'pre-inquiry,' they asked me why the Master's thesis hadn't been cited and asked nothing about the other paper at all. Richard Lynn, who was born in 1930 and was by that time was 86 and increasingly forgetful, submitted a document taking full responsibility for any errors in our supposedly plagiarized study, but this wasn't good enough for the complainants. It was argued that I must have known that the Master's thesis should have been cited and that I was responsible for the other study, as I was lead author. Lynn had come up with numbers from the military data involved without making the nature of his calculations clear. A few years later this intrepid, but very forgetful, retired academic had to have a target article (one which is followed by a series of critiques) withdrawn from a journal for self-plagiarism, it having been previously published elsewhere and Lynn having forgotten about this. The commentaries, which had already begun to be published, then had to be withdrawn (e.g. Cabeza de Baca & Woodley of Menie, 2017).

The responses of the complainants were accepted as needing an answer, so there was a formal enquiry. Three months later than I was told by a university lawyer this would occur, I was asked respond to what the complainants had said and given no deadline by which to do so. I explained to those investigating me – none of whom were native speakers of English – what I have outlined above. I also noted that the subject of the Master's thesis was the Flynn Effect and Richard Lynn had discovered this (Kanazawa, 2012, p.188). Why, therefore, would he deliberately and knowingly plagiarize a student thesis on this subject? Why would either of us do so when we had been widely published anyway and, in my case, I had already published in *Intelligence,* whose editorial board Lynn sat on? Why would we do so when the article would have been published in *Intelligence* even if the source had been made clear, as there are many examples of re-analyses of obscure unpublished theses being published? Why, moreover, would we draw attention to the fact that we knew about the existence of the thesis we were supposedly plagiarizing by actually citing it – as a source for information about Finnish army history – in the very article in which we were apparently plagiarizing it? Why would I, in a book of mine which went to press before we published this paper called *Religion and Intelligence* (Dutton, 2014, p.22), discuss this military data and state that it had been previously discussed in this Master's thesis? I stated that they had to decide between whether there had been a 'cock up' or a fantastically unlikely 'conspiracy' that anyone would realize would be uncovered. Surely, I told them, the simplest explanation is the cock up. I implied that you'd have to be schizophrenic to accept the latter as having occurred . . . but they did, stating that my explanation was, in their view, 'unbelievable' and that 'in all probability' I was guilty. This verdict was reached three months later than university regulations permitted; the rector having given them an extension without even telling me about it. I was given an opportunity to respond to the decision; my key points in this response were simply ignored, just as they had been in the first round. I was then given an opportunity to formally appeal to, as far as I could work out, a committee of the Finnish parliament. But in that this would be dominated by Leftists this seemed totally pointless. The university committee accepted that Lynn was solely to blame for issues with the 2014 study which I had not even been asked about in the pre-inquiry. I put a correction to the 2013 paper as the committee instructed.

Anyway, their verdict was gloatingly reported in the Finnish newspapers in June 2017. I couldn't quite believe what had happened. However, the more I thought about it the more it made sense in a society in which the average person is more schizotypal than is the case in many other European countries. It was also stated that the decision of the committee of investigation was 'unanimous.' This is almost always the case in Finland, for example, when a priest applies for a job and the decision is made by the parochial council. It almost certainly reflects little more than the Finnish desire to conform and the ease with which Finns will be dominated by the minority of zealous, assertive types. I suspect that my role in Finland as being known as a researcher of controversial issues may have played a role in my failure to receive the 'benefit of the doubt,' as people tend to be more paranoid and critical of those in their out-group (see Boyer, 2001). It wasn't a pleasant experience. But, on the plus-side, had it not occurred then I wouldn't have understood what it means to live in a society that is relatively high in schizotypy and I wouldn't have comprehended the significance of schizophrenia to making sense of the Finns.

4. Schizophrenia, Anxiety, and Life History Strategy

We shouldn't be at all surprised by the levels of schizophrenia in Finland. It is entirely consistent with their evolution to a cold yet harsh environment and this is further evidenced by the very high rates of schizophrenia in the wealthier Northeast Asian countries. In this regard, it should be noted that Finns – though not Finland-Swedes to the same degree – have around 10% Northeast Asian admixture (Virtaranta-Knowles et al., 1991) which is also likely to be relevant, as the ancestral environment of the Northeast Asians is very harsh yet stable. Indeed, as I observed in my academic article 'Battling to be European (Dutton, 2008), some linguists used to be convinced that Finnish was related to Mongolian and Japanese (e.g. Gleason, 1969) and it was commonly accepted, in the first half of the twentieth century, that Finns were not only non-Aryan (being Finno-Ugric) but also 'non-white,' an issue explored in depth, and with evident personal bitterness, in the book *Finns In the Shadow of the 'Aryans'* by Finnish historian Aira Kemilainen (1919-2006). According to mid-Nineteenth century racial taxonomies based on physical appearance, Finns were regarded as 'yellow' rather than 'white' (Virtanen, 1985). German anthropologist Johann Blumenbach (1752-1840) argued, based on skull shape, that Finns were 'nomads from the Steppes' (Hietala, 1985). The so-called 'Mongol Theory' of Finnish origins became orthodoxy (Kilpeläinen, 1985). Until as recently as the 1960s, encyclopedias in Nordic countries stated that Finns were 'mongoloid' (Aro, 1985, p.520). In the 1960s, among younger left wing Finns, there was even a renewed pride in their 'drops of Mongolian blood' (Kemiläinen, 2000, p.285). From the 1960s, the accepted 'narrative' began to change and Finnish promotional literature insisted that the country was as 'European' as any other (Anttonen, 2005). Kemiläinen (2000) stated, for example, that Finns were 'Old Europeans,' descendants of a particularly early settlement. Of course, this begs the question of why the Finnish X chromosome appears to have arrived in northern Europe from the east (Wiik, 2008). The issue of Finnish origins is contentious. But, as we have already discussed, the ecology of Northeast Asia – as well as of Northern Europe - would select for highly cooperative groups; groups in which members were relatively high in Agreeableness and high in Conscientiousness. In other words, it would select for people who had high impulse control and who were cooperative and non-aggressive. In doing so, it would select for those who have low levels of testosterone, because testosterone is associated with aggressiveness and low impulse control (Van der Linden et al., 2018). Autism is correlated with high levels of testosterone and those who are autistic tend towards being uncooperative and antisocial, a dimension of autism being low empathy (Baron-Cohen, 2002). Unsurprisingly, at the other end of the spectrum, schizophrenia is associated with low testosterone levels (Agarwal, 2013).

Paralleling this, left-handedness is associated with high testosterone, right-handedness is correlated with low testosterone and autistics are more likely to be left-handed while schizophrenics are more likely to be right-handed (James, 1988). Indeed, Finland has very low levels of left-handedness compared to other European countries – 4% in Finland compared to 12% in the UK - and left-handedness is low in Northeast Asia, being about 4% in Japan (Dutton, In Press). In other words, Finns are a low testosterone population. This would predict precisely the high Agreeableness and high Conscientiousness upon which foreigners in Finland remark. The high schizophrenia rate is simply a reflection of this. The ancestral ecology of the Finns would very strongly select for highly cooperative groups, meaning it would select against certain traits associated with autism and reduce testosterone (as a predictor of aggression and a correlate of autism) (Baron-Cohen, 2002). These cooperative traits would lead, by genetic chance or simply due to the average of the traits being higher, to higher levels of hypermentalism, the pathologically strong interest in the feelings of others that is the essence of schizophrenia. However, this would be an evolutionary price worth paying for the necessary extremely cooperative group (Dutton et al., In Press). So, it can be regarded as a component of a slow Life History Strategy.

If Finns follow such a strategy, one would expect them to be low in mental instability. However, as we have observed, they score higher on Neuroticism than the Finland-Swedes, which is unexpected. Similarly, Northeast Asians would be expected to follow a slow LHS, yet they are higher in measures of Neuroticism than are Europeans (Kura et al., 2015). It appears that this is because they are extremely high in social anxiety, to the extent that it overwhelms their low scores on maladaptive forms of Neuroticism, such as post-traumatic stress, and causes them, overall, to be higher in Neuroticism than Europeans or Africans (Fernandes et al., 2018). Social anxiety would be very important in a highly *K*-ecology, because it would make you far more concerned about offending people and avoiding social discord. It would thus contribute to creating a highly cooperative group. This is likely what underpins the high levels of Neuroticism among Finns.

Those who are socially anxious, as well as the highly intelligent and highly Agreeable (Nettle, 2007) tend to have low self-esteem. This is obvious in Finland if you are a foreigner because if you go to a bar and start speaking to a Finn they will often be interested to know what you think about Finland and the Finns. This concern about the opinion of others is a marker of high Agreeableness and low self-esteem, as it is people with these traits who are the most concerned about the feelings of others. Finnish sociologist Tarja Laine (2006, p.74) provides a detailed discussion of the Finnish inferiority complex. She claims that Finns are often 'discontented with their nation' and 'ashamed of themselves.' Laine argues that there is a history of 'self-stigmatisation' in Finland because Finnishness was defined as 'inferior to the rest of Europe' and the Finns were regarded as 'separate, mentally colonised "others."' Laine also notes how deeply concerned Finns were about what foreigners would think of them when they hosted the summer Olympics in 1952 and how there was a similar sense of national shame when a Finnish doping scandal came to light in the 1990s.

Myself and my colleague Guy Madison (Dutton & Madison, 2017) inadvertently provided evidence for the Finnish inferiority complex in a study we did of every marriage between a Finn and a foreigner that took place in Finland in the year 2013. On average, males and females operate different sexual selection strategies. Males have nothing to lose from the sexual encounter, so it makes sense for them, if they can get away with it, to have as much sex as possible with as many different women as possible in order to maximise the probability that their genes will be passed on. Accordingly, they select for youth and beauty, as these are markers of fertility and health. The essence of beauty is a symmetrical face and a such face implies a low level of mutant genes and thus sound genetic health. Females operate differently. As we discussed briefly earlier, they have a great deal to lose from the sexual encounter, because they can become pregnant, which carries with it a range of social and physical costs. This makes them more selective. Specifically, they are sexually attracted to high status men as these men will have the resources to provide for them and their child, meaning that both of them are more likely to survive (Buss, 1989). So, socioeconomically, women 'marry up' (*hypergamously*) and men 'marry down' (*hypogamously*). We would expect that nationality would be an aspect of status. We tested this by ranking different nationalities based on various criteria and especially how wealthy a country was. We predicted that, among marriages between a Finn and a foreigner, Finnish women would to a greater extent marry men that were from countries ranked as higher status than Finland while Finnish men would disproportionately marry women from lower status countries. This is, overall, what we found. However, we specifically found that, whatever the objective national status differences, Finnish women married Western European and Anglophone (USA, Canada and so on) men while Finnish men married Eastern European and East Asian (including Japanese) women. This would imply, whatever the economic reality, that Finns regard themselves as inferior to pretty much all Western Europeans. It also indicates that the Japanese – who are far wealthier than the Finns – regard themselves as inferior to the Finns, presumably because there is some idolization of whiteness or, possibly, as has been argued by a Japanese anthropologist, the Japanese specifically adore Finnish culture (Mitsui, 2012).

And returning to evidence of Finnish modal personality, Kura et al. (2015) have looked at the population prevalence of the DRD4 7-repeat allele. This has been shown to be associated with being of an inquisitive and questioning disposition and experiencing high levels of pleasure. In Europe, about 0.2% carry this allele, in Northeast Asian countries it is around 0% but in Finland, an extreme outlier in European terms, it is 0.061%. This would imply that Finns, like Northeast Asians, are genetically low in inquisitiveness and simply in feeling pleasure. This means they are less likely to 'rock the boat,' something which could potentially undermine the cohesiveness of the group. In line with this, based on assorted measures, Kura et al. note research indicating that Finns have the second lowest level of individualism in the West (based on 17 countries), with only Austria ranking – for some reason – as less individualistic than Finland.

5. Low Levels of Finnish Genius

Creativity levels in Finland are very low by European standards. Finns have a population of 5.5 million, not much different from Norway. We can all think of highly significant contributions to humanity from Norway: the lauded playwright Henrik Ibsen (1828-1906), the UK-raised children's author Roald Dahl (1916-1990), the composer Edvard Grieg (1843-1907), the painter Edvard Munch (1863-1944), Roald Amundson (1872-1928?) (first person to reach the South Pole), Tryggve Gran (1888-1980) (first flight across the North Sea), Thor Heyerdahl (1914-2002) (eminent explorer), Helge Ingstad (1899-2001) (explorer who proved there had been Viking settlements in America), Vilhelm Bjerknes (1862-1951) (father of modern weather forecasting), and Gerhard Armauer Hansen (1841-1912) (discovered the cause of leprosy). Coming up with a list of Finns who've made a very significant contribution to humanity is rather more difficult. There's the musician Jean Sibelius (1865-1957), the inventor of Linux Linus Torvalds (b. 1969), the people who made the computer game Angry Birds (perhaps that's rather stretching it, but it was very popular), Tove Jansson (1914-2001) (who wrote the Moomins), the philosopher Georg von Wright (1916-2003) and Marshal Carl Gustaf Mannerheim (1867-1951), internationally famous for leading Finland's plucky fight against the Soviets during the Winter War. All of these people – without exception – are or were Finland-Swedes. When it comes to the Finns, there's simply nobody to name; nobody who could, in any sense, be called a 'genius' or major player in the history of mankind.

Low levels of Finnish genius are most obvious in terms of science Nobel prizes per capita. Those who win science Nobel prizes have made a highly original breakthrough. As such, they are not just highly intelligent but also extremely creative. To a great extent, they are the essence of the 'genius' type and many accepted scientific geniuses have won Nobel prizes. As already discussed, despite Finns having an IQ that is the highest in Europe, their per capita Nobel prize achievement is extremely low. They have only won three science Nobels and there is a good argument for not counting two of those – those awarded to Ragnar Granit and Bengt Holmström - because they were both won by Finland-Swedes who were, also, no longer researching in Finland at the time. As such, the Finns are left with one science Nobel laureate: A.I. Virtanen (1895-1973) was awarded the Nobel Prize for Chemistry in 1945. This gives Finland 1.8 science Nobels per 10 million of the population. This is the same as Japan (IQ: 105) and only a little above Romania (IQ: 91) and Azerbaijan (IQ: 84.9).

The low levels of geniuses per capita among Finns can be neatly explained by a combination of everything we have looked at so far. A genius is a person who is widely recognised as having made a highly original and important breakthrough or other contribution to humanity (Simonton, 2003). Detailed examinations of the lives of accepted geniuses reveal that they are characterized by a combination of outlier high IQ and moderately low Agreeableness and low Conscientiousness, as we have discussed. In other words, they display certain traits associated with autism; they are high in testosterone. Indeed, my research group has found that when the analysis is limited to countries with an IQ of at least 90, then average national testosterone level robustly predicts per capita science Nobels. Congruous with this, there is a correlation of -0.55 between per capita science Nobels and national schizophrenia rate, -0.55 between per capita important scientific publications and schizophrenia rate, and 0.65 between science Nobels and the percentage of the population that is left-handed (Dutton et al., In Press). There is probably also something more specific to the personality profile that is associated with creativity, originality, and excellence because, as mentioned before, it requires the ability to think unconventionally, to withstand criticism and even ridicule, and to have the self-confidence and doggedness to pursue very hard work for many years with little or no support and sometimes even in the face of outright antagonism. That kind of person tends to be bold, critical, idiosyncratic, and unconventional, which is often perceived by people as him being annoying, arrogant, controversial, and provocative. Nobel laureates Norwegian neuroscientist Edvard Moser (b. 1962), American physiologist Brian Kobilka (b. 1955), American geneticist James Watson (b. 1928), and American physicist William Shockley (1910-1989) fit one or more of these characteristics, the two latter infamous for raising issues similar to the ones raised in this book. Regardless of whether genius is better conceived of as a particular trait or as a constellation of traits, it is clearly associated with masculinity and hence with level of testosterone. Indeed, almost all such people tend to be men, and it is therefore no surprise that this is true of science Nobel laureates in general (Van der Linden et al., 2018).

Finns, then, are low in creative achievement because they have relatively few intelligence outliers, relatively few personality outliers, relatively few antisocial people displaying certain autism traits, and so a relatively low percentage of people who combine outlier high IQ with these traits. This is why, if you watch Finnish television, almost everything on it that is 'Finnish' will actually be a

Finnish version of something British or American. Where TV producers in Britain, America or even Norway brainstorm original ideas, Finnish TV producers would seem to spend their time watching foreign programs to find things they can 'make a Finnish version of,' either overtly or covertly. Over the years I have been here, I have seen 'Finnish versions' of, among many others: *The X Factor, Idols, Master Chef, The Great British Bake Off, Have I Got News for You* (UK satirical news quiz), *The Apprentice, America's Next Top Model, Big Brother, Love Island, Ready Steady Cook* (even down to having 'red peppers' and 'green tomatoes' as teams), *Who Wants to be a Millionaire?, The Weakest Link, Long Lost Family* (where you find missing relatives; the Finnish version is interesting insomuch as the relative is always from abroad) . . . these are off the top of my head; there are likely many more. They struggle so much to fill the schedules that they even import Australian and Canadian versions of American or British programs, presumably because there are so many English-speaking subtitle producers: *Australian Princess* or the Canadian version of *Dragon's Den,* which is called *The Shark Tank.*

6. Northerners and Easterners

There are always exceptions to stereotypes about groups whether it's an extremely outgoing Japanese man or an intense and introverted Italian woman. So far, we have been discussing the stereotypical Finn and we have been exploring the most likely causes of this average Finnish psychology, in comparison to the psychology of other countries. Precisely because the modal Finn tends to possess this psychology, people who spend a relatively short amount of time in Finland will tend to notice this. However, as you spend longer in Finland you become more aware of regional personality differences which often parallel distinct regional accents. Stereotypically, Finns speak in a very monotonous way and they are low in Extraversion, high in Conscientiousness, high in Agreeableness, shy and not very talkative. The Finland-Swedes aren't like this but, genetically, they are a cline between the Finns and the Swedes so they are, in essence, a separate ethnicity. But even among the Finns you will notice stark differences.

Only today I was at the swimming pool in Oulu with my 7-year-old son. He, and another boy of his age, were happily engaged in throwing themselves off the diving board in assorted amusing ways: running along the board until it ran out; falling off the end as Charlie Chaplin might do if he was a 7-year-old boy from Northern Finland. I was sitting in one of the spectator seats at the side of the pool and so was the other boy's grandfather. He was expostulating various jokey instructions at his grandson about how to dive with greater hilarity. He seemed outgoing, confident; gregarious. And he definitely didn't have an Oulu accent. The way he spoke was too dramatic and mellifluous. Oddly, for someone in his 60s who was a retired car salesman, he wanted to talk to me in English, despite the fact that he spoke, in his words, 'only little' English. I hypothesized that he was either from Eastern Finland or Lapland and, in that there was still something just about reserved about him, I concluded it was more likely Lapland. I asked him where he was from. 'From here. Oulu,' he replied. I was surprised by this and asked if he had been born in Oulu. I was very pleased to discover that he had not been. Both of his parents were from Lapland and they'd moved to Oulu when he was four. Finns from Lapland, and especially Eastern Finns, do not easily conform to the Finnish stereotype. In October 2018, our family were returning from a holiday in Croatia and a middle-aged female taxi driver turned up at our airport hotel in Vantaa to take us to the airport. She was preposterously bubbly and talked at us for the entire journey. I asked where she was from just to be on the safe side but I needn't have. She was, of course, from Eastern Finland.

That these differences exist shouldn't be surprising but the reasons behind it are complicated. Genetic analyses have shown that Finns, in comparison to Swedes, have lower genetic diversity and a greater genetic affinity with East Asian populations. The Eastern Finns have a particularly small gene pool and an especially strong genetic relationship with East Asia (Salmela, 2012). Rick Kittles and his colleagues (Kittles et al., 1998) have shown that the Y-Chromosome of Eastern Finns is markedly different from that of Western Finns. If the Eastern Finns were the most genetically East Asian of Finnish populations then you might expect them to be the *most* stereotypically Finnish. But other evolutionary factors have complicated this neat and simple assumption. Until around the sixteenth century, Finns lived on the western and southern coasts. The interior wasn't really inhabited. At around this time, however, they began to migrate to the interior; to Lapland and to Central and

Interior-Eastern Finland (Kerminen et al., 2017). Most Eastern Finns descend from Finns from these southern coasts, these being populated and farmed while much of the interior was simply wilderness. In about the early sixteenth century, as stated, a small number of them migrated to the interior and then became isolated from the rest of the country for centuries (Taavitsainen, 2017, p.105). The founding populations of these interior Finns were so small and inbred that many genetic changes took place for reasons other than the well-known process of Darwinian selection. Most Finnish populations – but especially those in the East – display evidence, in their unusual genetic disease profile for example, of three specific forms of genetic selection: founder effect, genetic drift, and bottleneck effect (Salmela et al., 2008). Founder effect is a loss of genetic variation caused by a founding population being extremely small. Bottleneck effect occurs when there is a massive natural disaster which effectively kills at random, meaning that the survivors are not representative of the original population. Genetic drift is when the small size of the population means that particular genes simply disappear by chance due to certain people not reproducing. Accordingly, the high Extraversion of the Eastern Finns may reflect, for example, a small number of coastal Eastern Finns settling in the interior. There is some evidence, at least among males, that those who migrate are higher in Extraversion than those who do not (Tabor et al., 2015), which may be relevant to this process. Then, through these processes of chance, the Eastern Finns became the highly extravert people that we see today. The Central Finns are also products of this migration and, according to Kerminen et al. (2017), there is a genetic divide among the Finns between those who are from the coastal areas, that were settled earlier, and those who are from the interior areas, that were settled later.

The relatively outgoing nature of northern Finns is probably due to Saami admixture and it is clear from genetic analyses of Finns in a Lapland that they are indeed mixed with the Saami (Meinilä et al., 2001). Cochran and Harpending (2009) have argued that the development of agriculture would have elevated selection for such traits as Conscientiousness and Agreeableness. In that it would put a premium on carefully adhering to established systems, it would likely have selected against Extraversion (part of which involves risk-taking) as well. So, we would expect the Saami to be more outgoing than the Finns. We would also expect female Saami (inclined to marry hypergamously) to wed Finnish Lapland-based farmers and, thus, we would expect the Finns of Lapland to be relatively outgoing to the extent that they carry Saami genes. In addition, there is very low genetic diversity among the Northern Finns (Raitio et al., 2001). The Northern Finns, as with the Eastern Finns, made their way to Lapland from around the seventeenth century onwards and began to colonize it (Danver, 2015, p.615). So, the same processes that led to Eastern Finnish Extraversion may also have precipitated it among the Northern Finns.

However, this national psychology of few intelligence or personality outliers, high schizoid personality, low percentage of antisocial people with key autistic traits, and high social anxiety and conformism is generally what is prevalent in Finland, relative to other countries. And it has far more serious consequences than there being very few Finnish programs on Finnish television. It also helps to explain how Oulu's Muslim grooming crisis could carry on unchecked for so long.

Chapter Six

Finnish History, Kekkoslavia, and Democracy Under Pressure

1. Finland's Problem with Democracy

Before making sense of how this psychological profile has led to Finland's rapid Islamization as well as its suppression of reporting of the downsides of this process, it is worth exploring the kind of society which a national psychology of this kind would create. Actions speak louder than words, and an examination of Finnish history reveals exactly the kind of society which we would expect it to create: a society where you strongly conform and do not 'rock the boat;' a society in which there is a very high level of trust in those who rise to power, despite the strong evidence that those who tend to rise to positions of extreme power, such as leading businessmen and politicians, are generally highly intelligent psychopaths who are able to successfully manipulate less intelligent, more emotional or more trusting people (Post, 1994). But, as we will see, it is also a society in which individuals are prepared to make enormous sacrifices for the group and in which they are prepared to repel, with great bravery and even ferocity, a threatening out-group.

Finns generally believe that their country is strongly democratic, but 'Finnishness' is, in fact, not especially conducive to democracy. Democracy has been shown to be associated with a society's average IQ (Lynn & Vanhanen, 2012). This is because democracy requires organisation, the delaying of gratification, cooperation and trust; trust that if your opponents take power then they will faithfully maintain democracy, for example. Intelligence predicts not just believing in democracy, and so voting for democratic parties, but, specifically, democratic participation. This is probably because the more intelligent realise that if people don't participate in democracy then it will stagnate and the freedoms and stability which depend upon it will be lost (Deary et al., 2008). So, to put it very simply, democracy requires a certain level of average intelligence.

However, the correlation between national IQ and democracy is not especially strong: it is between 0.3 and 0.5 depending on what measure you employ (Lynn & Vanhanen, 2012, p.133). This is roughly the same as the correlation between intelligence and how much money you earn or between intelligence and how educated you are (Almlund et al., 2011). And the typical super-clever genius will often trade earning a large salary for being able to immerse himself in his ideas, while plenty of Conscientious people who aren't really that bright – the 'Head Girl'-types, as my colleague Bruce Charlton calls them (Dutton & Charlton, 2015) – may be highly educated and end up becoming professors of English Literature or Cultural Anthropology. Based on Lynn and Vanhanen's (2012) analysis, the relationship between a combination of measures of democracy and national IQ is 0.5. Finland is a strong positive outlier; it is *more* democratic than its IQ would predict. However, the problem here is that Lynn and Vanhanen seem to underestimate Finland's IQ and, moreover, as they note: 'On the basis of Vanhanen's Index of Power Resources (IPR) most of these countries are not large positive outliers because their high level of resource distribution predicts a high level of democratization' (p.147). In other words, the wealth of these countries interferes with our ability to understand how democratic they are because wealth also predicts being democratic. And, furthermore, there are certain very wealthy countries, such as Singapore, that are outliers in their lack of democracy. Indeed, in general, Northeast Asian countries are negative outliers in this respect.

I would argue that an optimum level of intelligence is necessary for sustaining democracy, in part because intelligence is associated with trust. Other factors also elevate trust, such as relative equality (Ulsuner, 2000) which, in Finland's case, is likely to partly reflect its narrow IQ and personality variability in personality. But if people are *too* trusting then they will be disinclined to hold their leaders to account, and this will damage democracy. Indeed, Jamal and Noordudin (2010) have shown that the relationship between democracy and generalized trust is quite complex and varies from country to country. They show that levels of generalized trust are linked to confidence in existing political institutions and, in democratic countries, these will tend to be democratic institutions. Accordingly, there is a tendency for high trust societies to simply trust their governments. If these governments slowly debase democratic institutions – such that it is not immediately perceptible that they are doing so – then it would follow that they would continue to be trusted and would not be challenged. So, too high a level of trust is bad for democracy. This, indeed, is consistent with the way that Northeast Asian countries, which we would expect to be very high in trust, are negative outliers in terms of their level of democracy. In line with this, Mark Warren (2018) has highlighted the fundamental 'paradox' at the heart of democracy:

'Democracy and trust have an essential but paradoxical relationship to one another. Democracies depend on trust among citizens, enabling them to depend upon one another. Trust in governments enables citizens to provide collectively conditions for good lives. Yet the institutions of democracy were founded on distrust, especially of the powerful.'

Democracy requires citizens, to some extent, to distrust the powerful; to question them and hold them to account. This, then, is Warren's paradox in action. As Eric Ulsuner (2000, p.6) has observed: 'People who trust others are less likely than mistrusters to endorse unconditional compliance. In the General Social Survey in the United States, just 35 percent of trusters say that you should always obey the law, even if it is unjust, compared to 48 percent of mistrusters.' So, the highly trusting are more likely to slavishly obey authority; to not question it. This is going to be a problem in a highly trusting society. To make matters worse, we would expect that it would be something close to the genius type – high in intelligence but low in Agreeableness (and so low in trust) – who would spearhead the process of questioning the current political dispensation and this is indeed the case (Post, 1994). A country with relatively few outliers would thus have relatively few people capable of performing this important task in maintaining democracy or, indeed, interested in instituting radical political change or even leading people, as great political leaders tend to display aspects of the genius type (Post, 1994). This problem can be seen in an examination of Finland's recent history.

It is clear that when threatened – at least when still strongly religious and relatively low in material wealth – Finns are relatively high in positive and negative ethnocentrism. In this sense, they are like the Japanese. They have adopted an 'ethnocentric' strategy because the nature of the ecology is such that a 'genius strategy' simply wouldn't work. Finland was a colony of Sweden – Finland was known as East Sweden – between the twelfth century and 1809, when it was ceded to the Tsar and became a Russian Grand Duchy. Throughout Finland's pre-nineteenth century history, Finnish regiments fought ferociously as part of Sweden's army and were known for their infamous battle cry of *Hakkaa päälle!* ('Cut them down!' or 'Strike on!') (Clements, 2014, p.30). Under Russian rule, many Finns fought courageously for the Tsar. In 1854, during the Crimean War, the British attempted to take various Finnish coastal towns but were invariably repelled by the Finns. When they attacked Kokkola, the British were not only driven away but one of their longboats was captured and remains on public display in a glass case in a central Kokkola park to this day (Jones, 1977, p.58). This extreme bravery makes sense, as we would expect the extremely harsh and extremely stable ecology of Finland to heavily select for strongly ethnocentric groups; internally cooperative and externally hostile. However, it is noteworthy – and we will see why shortly – that they were ultimately led by Swedes. Even so, this is a matter of standing-up to foreigners. As we will now see, Finns appear to be rather less prone to standing up to aggressors within what they see as their group.

2. National Awakening and the Awakening Movements

Under Russian rule, the Swedish-speaking elite, who made up the nobility and most of those who lived in cities, continued to rule the country. As late as the 1870s, Helsinki (then known, even internationally, as Helsingfors) was predominantly Swedish. The coast, and its cities, were, in most cases, majority Swedish while Finns were the majority in the interior and often changed languages if they became highly educated and so migrated to the towns (Talve, 1997). Throughout the nineteenth century, Finland underwent its 'National Awakening' in which Finns began to firmly see themselves, in the tradition of Romantic nationalism, as a distinct people (rather than a group of tribes) held together by language, blood, history and soil. However, this Finnish nationalism was, ironically, spearheaded by Romanticism-minded Swedish-speaking Finns such as Stockholm-born J.V. Snellman (1806-1881) and Elias Lönnrot (1802-1884), who made his way to Karelia to collect Finnish folklore from peasants and turned it into Finland's national epic *Kalevala*. As Karelia was the part of Finland least influenced by Sweden, it was regarded as a kind of pure, fossilised Finnish culture (Wilson, 1976). In this sense, even the innovation, or significant change, of the Finns adopting Romantic Nationalism was spearheaded by the Swedish-speaking population, as would be predicted by their evident higher per capita level of genius and, as implied by their lower Neuroticism, their lower levels of social anxiety.

Finland's 'National Awakening' developed at around the same time as a series of religious revivals which heralded the rise of so-called 'Awakening Groups' within the Finnish Lutheran Church. Most of these, however, were ultimately non-Finnish in origin. A prominent awakening group is the Laestadians. It has split into assorted subgroups and around Oulu, the so-called Conservative Laestadians are dominant and the term 'Laedstadians' now tends to refer exclusively to them. Fundamentalist in orientation, Conservative Laestadians traditionally reject television, make-up, ear-rings, the dying of hair, alcohol, and contraception, meaning that they tend to have very large families. Perhaps in response to these restrictions, they often smoke, dress very fashionably and the females style their hair remarkably meticulously. This movement was established by a half-Saami Swedish Lutheran priest called Lars Laestadius (1800-1861). Of the three other big awakening movements, two began among the Finland-Swedes (Larsen, 1998, p.490).

The third of these movements, which was in fact the earliest awakening movement, was led by a semi-literate farmer from Savo called Paavo Ruotsalainen (1777-1852). He travelled around Finland preaching and was even tried and fined, the authorities considering his activities so inflammatory. Followers of his piety movement would report visions and they would speak in tongues, shaking with religious ecstasy. Ruotsalainen's charisma, energy and tenacity should not be ignored, but nor should the fact that he was preaching a theology originated by his mentor, the Jyväskylä blacksmith and lay preacher Jakob Högman (1750-1806), originally from Ii, relatively close to Oulu. Högman was heavily influenced by the English theologian Thomas Wilcox (1621-1687) and his book *A Choice Drop of Honey* (Raunio, 2018). Thus, in effect, all Ruotsalainen did was spread Wilcox's views around Finland, Wilcox's books being 'the only spiritual texts Ruotsalainen valued beside the Bible' (Bruhn, 2003, p.305). So, the awakening movements were either led by Swedish-speakers or essentially inspired by an Englishman. But even so, it is clear that Ruotsalainen was a profoundly charismatic and rebellious figure; confident, prepared to make enemies, zealous. It is, therefore, likely no coincidence that he was from Eastern Finland where, as we have discussed, people tend to be higher in Extraversion and possibly lower in Conscientiousness.

3. The Finnish Civil War

The Finnish Civil War is also testimony to the Finnish psychology we have already discussed, though it is surely worth a short detour to understand how the war began. Under Russian rule, Finland had become an autonomous area with its own parliament representing the estates of the nobility, clergy, burghers and peasants. Russian rule had been relatively benign until 1899, when the Tsar began to introduce policies to Russify Finland. Russian was made the official language of administration, education, and the Finnish Senate (cabinet and supreme court). The Finnish army was abolished and Finnish conscripts were to be made to fight, alongside Russians, anywhere in the Russian Empire. As a result, at 11am on 16th June 1904, the Governor-General of Finland, Nikolay Bobrikov (1839-1904), was confronted in the Senate building by a Swedish-speaking Finnish nationalist called Eugen Schauman (1875-1904). Schauman, a civil servant and minor nobleman, shot Bobrikov, who died of his injuries in hospital that evening, and then shot himself, dying instantly (Sipilä, 2018). At the same time that Russia was Russifying Finland, Finland was being increasingly Finnicized by a '*Fennoman*' movement which was ultimately spearheaded by Romantic Finland-Swedes who learnt Finnish, insisted on using it, and, in some cases, changed their surnames to Finnish ones. It could be argued that, for Finland-Swedes, this identification with the rural and somehow purer Finns was a means of elevating their status, within the Finland-Swede community, through virtue-signalling.

The late nineteenth century witnessed the rise of Finnish-language schools and university education, something which mounted a direct challenge to the traditional dominance of Swedish. The result was a nationalist reaction by conservative '*Svenomans*' who attempted to preserve the status of the Swedish minority. During World War I, they collaborated with the Germans, leading to a 1,900-strong Finland-Swede battalion being trained in Germany. Meanwhile, the Social Democratic Party (SDP) was established in 1899 to represent the growing industrial working class in the large cities, which were mainly in the south of Finland. By 1904, there was a General Strike in Finland which the Tsar dealt with by establishing, in 1906, a parliament with universal suffrage. The Social Democrats had a majority in this parliament, which the Tsar neutralised by imposing a multi-party Senate. In February 1917, the Tsar was deposed, the police force was abolished, the army fell apart, and Russia collapsed into anarchy. In the Finnish elections of October 1917, the SDP lost their majority, but, by that time, the power vacuum caused by the collapse in public order had led to the setting up of various paramilitary forces to preserve the rule of law, roughly divisible into the 'White Guards' and the 'Red Guards.' This meant that the Whites controlled some parts of Finland while the Reds controlled others, specifically those where SDP support was strong.

The October Revolution took place on 7th November 1917 (still October on the Julian calendar that Russia observed), meaning that Russia had fallen to the Communists, though a civil war raged between 1918 and 1922 (Bullock, 2014). The Finnish parliament, in which the White forces had a majority, declared themselves sovereign and declared independence. In essence, the Finnish government, substantially composed of Finland-Swedes, who even now compose 50% of the Finnish nobility (Dutton, 2009b, p.180), broke away and declared independence for the small country of just 3 million. It was the Finland-Swede elite, in the form of Baron Carl Gustaf Mannerheim (1867-1951) and his allies, who made this daring move, with Mannerheim acting as state regent of Finland from 12th December 1918 (6 days after independence was declared) to 26th July 1919. Finland's Declaration of Independence was accepted by Lenin, who had his own power seizure in Russia to concentrate on, on 18th December (Service, 1995, p.285). This declaration was not accepted by the Social Democratic Party, the extremists in which were supporters of Lenin. They called a General Strike, became ever more radicalised, and, by this time, controlled Finland's industrial cities, including Oulu, but especially the more industrialized south. Accordingly, there were now two *de facto* states in Finland: White Finland, run by Mannerheim and his colleagues, and Red Finland, known as 'The Finnish People's Delegation,' led by the Social Democrat Kullervo Manner (1880-1939) who, as we would predict based on his propensity to spearhead radical change and conflict, was from a Swedish-speaking background. The other key Red leader, the head of the Red Guard, was Eero Haapalainen

(1880-1937). He was from Kuopio, in the East of Finland. An arms race commenced, and by 27th January 1918 there was full-scale civil war. This extraordinarily bloody conflict lasted until 18th May 1918. Approximately 6000 Whites and 32,000 Reds were killed. In just 3 months, therefore, over 10% of the Finnish population were slain. Many of the leaders of the Finnish People's Delegation promptly fled to the Soviet Union. Haapalainen was executed during the 'Great Purge' of 1937.

This savage whirlwind of bloodletting touched every Finnish family. My Finnish father-in-law, the Very Rev. ('*Rovasti*') Matti Hänninen (1940-2016), used to be leader of the Social Democrats on Kokkola Council. I once asked him why he had become involved with the SDP. He told me that when he was about 12 he had discovered that his maternal grandfather, a baker called Edvard Ammesmäki (1893-1918), had been summarily executed near Kokkola (on 6th April 1918) as a captured Red. My father-in-law's mother, Elsé Ammesmäki (1918-1989), who had been born on 19th January 1918, was duly raised by her maternal grandparents, who were Finland-Swedes. This meant that her native language was Swedish, though, due to the Finnish nationalist atmosphere of the time, she never spoke this language to her children. My father-in-law was so horrified by the discovery of what had happened to his grandfather – of what was a deeply shameful family secret that was never discussed – that he became drawn towards the Social Democrats, as if in honour of his martyr grandparent. My mother-in-law's mother, Hilkka Träskman (1910-1999), later a school teacher, was raised in Tampere. Her father, a Red, was interned in Tampere and she recalled walking past corpses in the street as the family made their way to the prison camp to give her father food.

In the wake of the Whites' victory, a reign of White Terror began. Scores of thousands of defeated Reds and their families attempted to flee to the Soviet Union. Some, including much of the Red leadership, were successful, but most were rounded up by the White Guards. Around 75,000 Reds – including women and children – were herded into concentration camps. In these squalid prisons, 13,000 died simply of epidemics and malnutrition (Lavery, 2006, p.87). Approximately 8000 – again, including women and children – were hanged or shot as traitors (Trotter, 2013). So, as far as I can see, the Civil War can be seen as a manifestation of Finland's very strong ethnocentrism – in the broadest sense of the word; a propensity to sacrifice for the in-group and abominate the out-group, at least under sufficient stress – combined with a tendency to allow themselves to be led by Finland-Swedes or even Eastern Finns, though it should be stressed that many significant figures – such as sometime Red Guard leader Ali Aaltonen (1884-1918) – were from the west of the country. The country split into two factions both of whom regarded the other faction as the out-group – almost as 'foreign' – and they were prepared to act lethally accordingly.

4. Finnish Nationalism and Democracy

In the wake of the Civil War, Finland was established as a democracy but – seemingly as a result of high levels of trust and a lack of a desire to stir things up – this 'democracy' quickly became debased and was very much 'in name only.' The fiercely nationalist Lapua Movement was set up in 1929 by various former White Guards. This nakedly violent group protested against the right of the Communist Party to operate in Finland and, in the summer of 1930, smashed up the party's newspaper presses in Oulu and Vaasa, beating up former Red Guards in the street (Jussilä et al., 1999). They also objected to any criticism of Mannerheim, the White Guard leader, who, like them, espoused a 'Greater Finland' in which all of the Finnic-language-speaking parts of the Soviet Union, such as Eastern Karelia, would be one great nation (Nenye et al., 2015, p.33). In the summer of 1930, under the leadership of former White Guard Vihtori Kosola (1884-1936), the Lapua Movement abducted two socialist parliamentarians from parliament itself and only released them when every Communist MP had been arrested. It also led a 12,000-person Anti-Communist march in Helsinki. In elections in October, the threat of Lapua Movement violence led President Lauri Relander (1883-1942) to advise Finns to vote for parties that would introduce the anti-Communist laws demanded by the Lapua Movement. They duly did so, appeasing the aggressor in the process. Regional governors were allowed to effectively stop socialists from electioneering. To avoid further violence, the right wing winners of this sham election simply caved into pressure from these anti-democratic forces and outlawed the Communist Party's newspaper, also placing other restrictions upon the Left.

In effect, by 1930, the Lapua Movement dominated the government and the country. Lapua Movement activists shut down left wing meetings. They even kidnapped various left wing activists, drove them to the Russian border and forcefully deported them into the Soviet Union. This resulted in many of these Reds being executed as 'nationalists' during the Great Purge of 1937, by virtue of their leadership of the then semi-autonomous Soviet Republic of Karelia, which was, at that time, overwhelmingly Finnish-speaking. In October 1930, the Lapua Movement went so far as to kidnap the former president, Kaarlo Ståhlberg (1865-1952), and drive him to Joensuu (close to the Russian border), presumably in order to display their power. In late 1931, parliament was intimidated into electing the man the Lapua Movement supported, Pehr Evind Svinhufvud (1861-1944), as president. The election of this so-called 'untitled nobleman' – a minor nobleman, roughly equivalent to the English armigerous gentry or 'esquires' (see Dutton, 2015, Ch. 2) - occurred amid rumours of an impending coup by the Lapua Movement. Lauri Malmerg (1888-1948), a former White Guard, telephoned the leader of the Agrarian Party to make it clear that if former president Kaarlo Ståhlberg (who coincidentally was raised in Oulu), another of the candidates, was elected then Malmerg couldn't guarantee order. A right wing newspaper published a thinly veiled incitement to assassinate Ståhlberg should he be elected and 'armed Lapua men were already in the capital ready to unleash a bloodbath if Svinhufvud were not elected.' The majority of Finns had clearly voted for parties opposed to the Lapua Movement. Svinhufvud's bloc constituted just 21.6% of the vote while the Social Democrats commanded 30.2%. But MPs, who constituted the electoral college, elected the Lapua Movement's desired candidate anyway (Kirby, 1980, p.89). This was a kind of *de facto* coup. The Lapua Movement, it could at least be argued, were effectively running the country.

The Lapua Movement were outlawed after a failed coup – the Mäntsälä Rebellion – in November 1932. However, the coup leaders, such as General Kurt Wallenius (1893-1984), received relatively light sentences (Jussila et al., 1999). Indeed, General Wallenius was back commanding soldiers by the time of the Winter War in 1939. President Svinhufvud ensured that the Social Democrats, despite being the largest party, were kept out of government for the length of his presidency, which ended in 1937. He has been described as the 'champion' of the Lapua Movement (Barnett, 2007, p.333). We can conclude from what occurred in Finland in the 1930s, that there simply wasn't the will to maintain democracy in any meaningful sense. The Finnish electorate were prepared to vote for parties that would persecute the Left – and thus suppress democracy – as long as it meant an end to political violence; as long as it meant a quiet life. The Finnish parliament was prepared to elect as president the man whom the Lapua Movement commanded for the same kinds of reasons: a highly Agreeable people appeased rather than stood up to a dictatorial aggressor from, within its own ranks.

5. The Winter and Continuation Wars

The two sides of Finnish ethnocentrism can be starkly observed in the country's relationship with Russia. On the one hand, the country's period of war with the Soviet Union between 1939 and 1944 inspired many heroic acts, a particular hero of the Winter War being the sniper Corporal Simo Häyhä (1905-2002) who accrued 542 confirmed kills, often taking considerable risks to achieve his goals and being seriously injured in the process (Saarelainen, 2016). But on the other it only heightened a deep hatred of Russians which exists to this day. As one Finnish writer has put it: 'When I was a child it was normal to be racist especially against Russians and gypsies, as long as you were racist behind their back. They had to be ignored and made to feel unwelcome in Finland. My grandparents, who had been in the war with the Russians together with their Nazi German allies during the Second World War, told me over and over again: "Russians are Russians even if you stir fry them in butter"' (Kannisto & Kannisto, 2012). In November 1939, Stalin demanded that Finland cede Eastern Karelia to the Soviet Union. The Finns wouldn't do so and, thus, the Soviets invaded on 30th November. In what became known as the Winter War, the Finns held out for a remarkable amount of time, before they were finally overwhelmed by the sheer numbers of Soviet troops in March 1940. Finns who wished to be, such as Tatu Vanhanen whom we met earlier, were evacuated to within Finland's new borders. On 25th June 1940, the Finns began the Continuation War, in which they retook Eastern Karelia and then further parts of Karelia that had never been part of the Finnish state. The Finnish Karelians, Vanhanen's family among them, mostly returned to Eastern Karelia (Dutton, 2015a). There were various motivations for this, in particular that Finnish leadership, such as Mannerheim and its president Risto Ryti (1889-

1956), wanted to create a Greater Finland, encompassing all areas where Finnic languages were spoken. To maintain support for this war, which the SDP opposed, Ryti brought the IKL, the successors to the Lapua Movement, into the government.

Ultimately, in September 1944, the Finns were forced to permanently cede Eastern Karelia, and it was evacuated of Finns again. They also had to cede parts of Lapland, meaning that they lost 12% of their territory including Viborg, their second largest city (see Wuorinen, 2015). However, when the Finns retook Eastern Karelia in 1940, Mannerheim was explicit that any Russians there should be placed in concentration camps, as were Finnish leftist agitators. By 1942, 27% of those living in Eastern Karelia were in concentration camps (Gullberg, 2011). Those who were racially Finnic – such as Finns, Estonians and Karelians – were free and received sufficient food rations. The Russians were put into concentration camps where they were given insufficient rations. These camps had a very high mortality rate: 17% of those who went in died there due to malnutrition and disease (Silvennoinen, 2011, pp.387-389). Captured Soviet soldiers were put into labour camps, where they had to work bare foot, and were given starvation rations. By the summer of 1942, many were reduced to eating grass. Approximately, 1,200 of these POWs were shot without trial. These were 'illegal executions of prisoners, often committed out of a variety of causes including fear, hatred, incompetence, alcoholism, and simple sadism' (Silvennoinen, 2011, p.378). I would argue that these camps are a testimony to relatively high levels of Finnish negative ethnocentrism.

One extraordinary example of the positive ethnocentrism of the Finns during this period is that the government mandated, and the population accepted, that people must divide up their properties in order to take families of Karelian refugees into their homes. A room rationing system was imposed, on the basis of one family per room (Insall & Salmon, 2011, p.112). My father-in-law, whom I mentioned earlier, recalled precisely this happening, even though their house wasn't particularly big. It was divided down the middle, with Karelians living in one half and his family squashed together in the other. Farmers also found parts of their land subject to compulsory purchase so that Karelians could farm the land instead (Kukkonen, 1969, p.40).

6. Between East and West

However, in the wake of the War, a new system of thinking – and then a particular Eastern Finnish autocrat, the wily and prematurely bald Urho Kekkonen (1900-1986) – were able to debase Finnish democracy yet again, with relatively little opposition. Finland's Post-War policy became one of appeasement to the Soviet Union. The Paasikivi-Kekkonen line, named after the two successive presidents that pursued it, involved Finland cooperating with its giant neighbour while also cooperating with the West. This cooperation with the Soviet Union, and, in effect, willingness to allow it to tacitly influence Finnish affairs, became known as 'Finlandization.' In 1944, in order to maintain peace with the Soviets, Finland banned assorted nationalist organizations, including the Greater Finland-promoting Academic Karelia Society. In 1956, a former member of this society and former White Guard, Urho Kekkonen, of the Agrarian League (later the Centre Party) was elected president by the country's electoral college. This lawyer, journalist and Finnish high jump champion had a magnetic personality. Through his charisma, he was able to persuade people that there was an imminent threat of Soviet invasion and that only he had the skill to successfully negotiate with the Soviet leadership and save the country. With his large following, people in positions of power were prepared to do his bidding. If you questioned him – and particularly his foreign policy – he would declare you to be 'out of favour at court' (Austin, 1996, p.2). This could then have serious consequences for issues such as employment.

Under Kekkonen's lengthy rule, which lasted until 1982, Finland was perceived to be so co-operative with the Soviets as to no longer really be independent and, indeed, Finnish historian Seppo Hentilä (2001) has stressed that in an international crisis, Finland would have been in the Soviet sphere of influence. Unlike the rest of this sphere, Finland remained a multiparty democracy, but by the 1970s, Kekkonen's power was so great, his control of the media and the parliament (who elected him) so tight, and censorship so endemic (in fact strong criticism of the Soviets was effectively illegal) that the reality of Finland's 'democracy' came into question in the West. National policy was, in many cases, okayed by Moscow before being put into action. No wonder Leonid Brehznev quipped in 1973, 'Finland is in the back pocket of the Soviet Union.'

As early as 1992, politician Risto Pentillä (1992) was writing that Kekkonen created an 'official religion' of Soviet kow-towing during the 1970s, he allowed Finland to essentially lose its independence and integrity: 'His powers resembled those of an old-fashioned monarch.' The trusting Finns – so goes the narrative – were betrayed by one of their own. Kekkonen went way beyond what was required to maintain independence. He played the Moscow card to preserve his power, something that would be highly effective in a relatively schizoid society. Parties that criticised him, such as the conservatives, were, most of the time, kept out of the governments which he appointed. Since the Cold War ended, historians have presented more and more evidence of Kekkonen's Machiavellian machinations to substantiate this narrative. In 1961, it was looking like Kekkonen might lose the presidential election to conservative Olavi Honka (1894-1988). The USSR sent a 'note' referring to the threat of 'war,' possibly because they wanted Kekkonen re-elected. This created the 'Note Crisis,' Kekkonen went to the USSR to sort it out and eventually he was re-elected overwhelmingly after Honka withdrew. It is argued that Kekkonen manipulated and planned all of this in order to ensure his re-election. In 1973, he gained support for an emergency law – by implying that, if he were not re-elected, the Soviets would oppose Finnish membership of the EEC - so that he could carry on being president without an election (Dutton, October 2009). Thus, for the second time, the first being due to the Lapua Movement, Finland had effectively lost its democracy, but on neither occasion did it manage to do so in a black-and-white way, such that it could unarguably be termed a 'dictatorship.' Historian and journalist Jukka Relander argued in 2001 that the Finns have 'always bowed down before external powers.' First it was the Swedes, then the 'Lutheran God,' then the Russians, then the Germans (there was much German cultural influence in Finland after independence), then the Soviet Union and now 'International Capitalism' and the European Union. Finlandization is in the depths of the Finnish psyche. 'Finland does not decide; it reacts' summarises Relander. It sees itself as 'a victim' (Relander, 2001). In this case, it was, once more, a victim of a tendency not to 'stir things up' by causing conflict *within* the group

.

7. Joining the European Union

So, again, with Kekkonen, we see how difficult it is for the Finns to maintain democracy. They are too inclined to follow and trust their leaders, they are disinclined to stand out from the crowd and risk social opprobrium, or, rather, there are too few per capita people who are prepared to behave in such a way or support those who are prepared to. Moreover, it could be argued that Kekkonen successfully took advantage of a kind of paranoia among the Finns. As we have discussed, they are relatively high in schizophrenia, meaning that the average Finn is further along the schizotypy spectrum than is the average person in many European countries. This would mean that a higher proportion of Finns, with their very high empathy, would read too much into the external signs of the mind of the Soviet Union and thus become paranoid, prepared to assume that an indication of displeasure was in fact an indication of fury, possibly leading to invasion. 'Only President Kekkonen can deal with this crisis' they might reason, 'so I must support him.' But, in reality, there isn't really a crisis at all.

The same democratically problematic tendencies could be observed when Finland voted to join the European Union. Seppo Hentilä (2001) while arguing that Finland was 'democratic' under Kekkonen, nevertheless refers to an 'almost compulsory national consensus' during this period. He also claims that many Finns looked upon the country's referendum on joining the EU as a 'national crisis' that might tear the country apart, so unfamiliar were Finnish people with the concept of the public expression of a plurality of opinions. According to Relander, when the country voted to join the EU in 1994, it asked itself not if joining the EU was beneficial but, 'How can we make ourselves fit for the EU?' Joining the EU, for Relander, was another chapter in this process of Finlandization, of intense concern about the perceptions of others, of national insecurity and kow-towing to those perceived as important (see Dutton, October 2009). The Finnish elite wanted Finland to join the EU and so they strongly appealed to the fears of the Finnish public, a mere 3 years after the collapse of the Soviet Union, with regard to national security. Ultimately, 57% of Finns who turned out voted to join, 26% didn't vote at all, and it was the pro-EU, and more heavily populated south, that pulled the anti-EU north into the

B|E|C|U

LOCATION: 100 NW 85TH ST.
SEATTLE   WA

CARD NO: XXXXXXXXXXXX7096

DATE	TIME	TERMINAL
09/12/19	11:33AM	WA033404

A0000000042203
DEBIT

SEQ NBR: 3655   AMT: $240.00
CHECK DEP - CHECKING
AVAILABLE BALANCE $32,587.48
BALANCE $32,587.48
    Item 01 $240.00
    THANK YOU FOR USING THIS BECU ATM

European Union (Raunio & Tiilikainen, 2004): Overly high trust, over-analysing, and not rocking the boat had struck again.

8. Urho Kekkonen, Mutants and Multiculturalism

Urho Kekkonen was an autocrat. However, he was an ardent Finnish nationalist, certainly in his youth. Sweden had become a socialist country in the 1930s and developed into what has been termed a 'soft totalitarian' society in which, in effect, socialism and, increasingly by the 1970s, its virtue-signalling successor of Multiculturalism simply couldn't be questioned in public. The Social Democratic Party managed to politicise, and take control of, almost every aspect of Swedish society, including the judiciary, which was run by a network of Social Democrats. Sweden saw itself as a uniquely egalitarian, equal, and Leftist place and to think otherwise was un-Swedish. By 1971, Social Democrats had run the government for 40 years and a huge and powerful bureaucracy had developed in which Social Democrats would, of course, appoint only those whom they deemed ideologically sound. Thus, in a theoretically multiparty state soft totalitarianism was able to take hold (Huntford, 1971).

Kekkonen consolidated his own power through similarly devious methods. But unlike those who ruled Sweden – such as Social Democrat Tage Erlander (1901-1985), Prime Minister from 1946 to 1969 - Kekkonen evidently had strong nationalist sympathies, though they were merely *sympathies*. Kekkonen resigned from the Academic Karelia Society in 1932 over its support for the Mäntsälä Rebellion and as Justice Minister in 1936 he attempted, unsuccessfully, to ban the ultranationalist Patriotic People's Movement (*Isänmaallinen Kansanliike*, IKL) which was a successor to the Lapua Movement. This political party operated a paramilitary group known as the Blue-Blacks and, once they were banned, as the Black Shirts. Led by a Lutheran priest called Elias Simojoki (1899-1940), who was later killed fighting in the Winter War, members even used the Roman salute. Coincidentally, a future Bishop of Oulu (1963-1965) and Conservative Laestadian Pekka Tappinen (1893-1982) was an IKL Member of Parliament between 1933 and 1935. Historian of Finland Max Jakobson (1987, p.107) has summarised that Kekkonen, though he regarded himself as a 'champion of the poor against the privileged' was 'a nationalist not a socialist.' Indeed, in 1932, when he was already an MP, he reportedly said, '"Never mix with a yid if you can be among Christians," that is my motto' (Muir, 2013). A seemingly affable character, based on how he interacted with people, under Kekkonen's rule Finland became a wealthy country for the first time and it also developed a generous welfare state (Kirby, 2006, p.275). But he can be regarded, nevertheless, as a nationalist, and, from his perspective, Finland's Cold War neutrality was motivated by a desire to allow the Finnish nation, as then constituted, to continue existing (Bergmann, 2016, p.80). To a certain extent, Kekkonen's quarter-decade rule kept Finland sealed off; protected it from the anti-nationalism and Multiculturalism that was affecting many Western countries, especially by the 1970s. Kekkonen, however, couldn't do much about the rejuvenation of Marxism in Finland – especially considering the delicate relationship with the Soviets – and these ideological successors to the defeated Reds were, by the 1990s, espousing environmentalism and Multiculturalism. Any film that is now made in Finland about the Civil War is to some degree pro-Red. Examples include *Raja-1918* (*The Border,* 2007) and *Käsky* (*Tears of April,* 2008). But, in his defence, Kekkonen helped to delay Finland's enrichment by Multiculturalism, giving it the possibility of doing something about that enrichment earlier on in the process than would be the case, for example, in Sweden.

It seems likely that Finland's specific evolutionary context also helps to explain its delay, relative to Western countries further south, in embracing Multiculturalism. Returning to the Social Epistasis Amplification Model, Finland industrialized considerably later than much of Western Europe and had higher levels of extreme poverty until much more recently (see Kirby, 2006). This would have meant that it would take until a later generation, than was the case in Britain, for the country to be sufficiently high in spiteful mutants for the process of societal collapse to be set off, likely best expressed in the undermining of the power of the Finnish Lutheran Church. So this would explain why the process took so long to reach Finland. But once it began, we would expect it to happen extremely quickly. This is because Finland is such a strongly cooperative society, meaning that the different parts of the whole are mutually influenced by each other to a very strong extent. Thus, the collapse of the traditional Finland would occur at breath-taking speed compared to the situation in many other European countries. And this has likely been further accelerated by the fact that the Finnish political leadership has long been relatively young, meaning even higher average levels of spiteful mutations. Three of the four Prime Ministers who have been in office since 2010, for example, were well-under 50 when they took power and one of them, Jyrki Katainen, was just 39 years-old. Jutta Urpilainen, the leader of the Social Democrats between 2008 and 2014, was only 32 when she took up the influential post. She is seemingly infertile – a marker of high mutational load – as she and her husband adopted a Colombian baby in 2016.

9. The Enrichment of Suomi

But the enrichment occurred, as did the consequent virtue-signalling and attempts to stifle freedom of expression by the Multicultural zealots. Finnish folklorist Pertti Anttonen (2005, p.125) observed in 2005 that:

> 'According to recent statistics and polls a large portion of the population in Finland has a somewhat negative attitude towards the recent growth of immigration . . . Approximately 20 percent of the 1000 persons interviewed (for a 1998 poll) favoured the view that people from different cultures should not mix.'

In the year 2000, it was reported that 42% of Finns felt that certain races were unsuitable to live in Finland (Anttonen, 2005, p.125). As noted above, in 2004 the police publicly announced that they were considering investigating Prof Tatu Vanhanen (1929-2015) (the then Finnish Prime Minister Matti Vanhanen's father) because he gave an interview to a magazine about his peer-reviewed research *IQ and the Wealth of Nations* (Lynn & Vanhanen 2002) which argued that Japanese people have the highest IQs, black people have the lowest and national IQ is directly reflected in the wealth of nations (*Helsingin Sanomat,* 12th October 2004). Ultimately, Vanhanen was not 'investigated' for summarising his scientific research. In late 2005, a Danish newspaper published satirical cartoons of the Prophet Mohammed and Norwegian newspapers republished them out of sympathy for the cartoonists, in the wake of the rioting and furore that they caused among offended Muslims. Norwegian and Danish embassies were burned down in the Middle East and there was a riot outside the Danish Embassy in London. The Norwegian and Danish governments also openly supported the right of the newspapers to publish the cartoons. Finnish newspapers did not publish them, despite Finland having no sizeable Muslim minority. When the website of a small nationalist group – *Suomen Sisu* – did publish then, the Prime Minister and President publicly apologised for this (*Helsingin Sanomat*, 16th February 2006) and unsuccessful attempts were made to prosecute the group (*Helsingin Sanomat*, 26th June 2006).

By the time I came to Finland, the attitude of the kind of people who work for Oulu City Council – and live in detached houses far from the city centre, meaning they don't have to live with the consequences of Multiculturalism – was that Oulu should 'internationalise' and become more 'Multicultural.' This was regarded as good for the economy, a good in itself, and also as a way of turning Finland into a proper, 'modern' Western European country, just like the Multicultural 'big boys,' such as Sweden and the UK. Indeed, this seemed to coincide with the Finnish elite altering the narrative of what it even meant to be Finnish. As I noted in *The Finnuit* (2009), Pertti Anttonen (2005) has charted a process of 'Westernizing' in how Finland is presented in official books aimed at foreigners. The idea that Finns have Eastern origins has been gradually removed as has the idea that they are 'between East and West.' By the time he did his research they were entirely 'Western' and 'Multicultural,' which, if they so obviously are, why should it even be asserted? Finland, however, was now to be entirely 'Western' - notwithstanding the fact that my father-in-law had been taught at school that Finns were from the East and were, to some degree, non-white – and Multicultural. In May 2008, some department of the council went so far as to organise a 'Multicultural Mother's Day' in which refugees were invited to perform various national dances for the edification of the 'woke' Finnish, upper-middle class audience.

In 2009, Dr Jussi Halla-Aho, now an MP and leader of the party True Finns (*Perussuomalaiset*), was convicted of 'disturbing religious worship and ethnic agitation' and ordered to pay a fie of 330 euros. His crime was to have satirized a comment made by a writer for the Oulu newspaper *Kaleva*. This writer had asserted that drinking heavily and killing when drunk were cultural and possibly genetic characteristics of Finns. Halla-Aho had retorted, on his blog that, by the same reasoning, robbing people and living off welfare were cultural and genetic characteristics of Somalis. For this, he was prosecuted and convicted. It leaves one utterly speechless; more speechless even than a Finnish stereotype. One can only hope that one day the anti-free-speech advocates who played any role whatsoever in prosecuting or convicting him will receive justice. Only two years later, in 2011, Halla-Aho was swept into parliament, as True Finns (whose 'official' English name is now 'The Finns') went from 4% to 19% of the vote, making them the third largest party and pushing the governing Centre Party into fourth place. Shortly after this, one of True Finns' new MPs, a saw mill owner called Teuvo Hakkarainen, did an impression of the Muslim call to prayer in an interview and used the word *neekeriukko* ('nigger uncle'). For this potential speech-crime the Ombundsman for Minorities investigated whether Hakkarainen could be prosecuted (*Helsingin Sanomat,* 11th May 2011). In the 2015 election, True Finns garnered 17%, making them the second largest party. They joined the government, led by Centre Party Prime Minister Juha Sipilä.

10. The Collapse of the Finnish Lutheran Church

During my time in Finland, the rise of Multiculturalism has, predictably, run in parallel with the collapse of the Finnish Lutheran Church. In Finland, there is a system of church membership where members pay a small 'church tax' as a percentage of their salary. However, the church's membership is falling at an ever-faster rate. In 1980, 90% of Finns were members. By 1990, it was down to 88%, in the year 2000 it was down to 85%, by 2010 membership was at 78% (Taira, 28th November 2015) and, at the time of writing, it is 69% (EVL, 28th January 2019). Since about 2016, Oulu schools have been increasingly abolishing the tradition of *pikkukirkko,* where teachers take their classes to church, as well as the tradition of asking priests to come and speak at their schools. In some cases, 'woke' teachers have demanded, for example, that the priest come and speak at the school but not say anything about God or even that there should be a *pikkukirkko* but could the priest please avoid any mention of religion. In 2014, I was asked to teach a course on Religious Studies to Education students at Oulu University. I was aghast by how rabidly anti-Lutheran one of the (female) students was. This slender, 20- year-old blonde smoker argued, in her essay, that she wanted to abolish religious education in schools (which pupils receive only if their parents are church members) and replace it with Multicultural education, which would be 'compulsory' for everyone and would teach them about racial equality and how there's no God.

The vast majority of Finns attend a week-long 'confirmation camp' aged about 15, rendering them emotionally attached to the Lutheran Church and the Finnish nationalism of which it was so significant a part: 'a good Finn is also a Lutheran' (Müller, 2016, p.47). They then have a big party to celebrate their confirmation and receive lots of money and presents from relatives. Having been confirmed and raised with the Lutheran Church has, therefore, been a significant part of Finns' lives, especially in the north and in the countryside. Accordingly, many Finns need an 'excuse' to 'divorce' the Church. In October 2010, the Finnish Interior Minister, Dr Päivi Räsänen, who was leader of the small, fundamentalist Christian Democrats party, stated in a televised debate that homosexuality was a 'sin.' Though she held no position in the Church, people took this out on the Church and there was a huge spike in resignations from it. More than 36,000 people left the Church in just two weeks (*Church Times,* 27th October 2010). The same thing happened, though on a much less dramatic scale, in 2013 when Räsänen expressed her distaste for abortion (*YLE,* 11th July 2013). On a personal note, I should add that I interviewed Päivi Räsänan in October 2010 for the newspaper for which I worked. She didn't answer the phone but eventually called me back, despite not knowing who had called her. Speaking excellent English, she seemed very reasonable and perceptive. She calmly explained to me that: 'Marriage is an agreement between a man and a woman. This is a Christian view . . . I see it in the Old and New Testament and in Jesus' teaching.'

In 2014, the Archbishop of Finland, the Most Rev. Dr Kari Mäkinen, firmly declared his support for gay marriage (Dutton, 5th December 2014). In 2017 a Hungarian-speaking Serbian priest based in Oulu – the Rev. Arpad Kovacs - in the Church, against Church Law, performed a gay marriage (Ukkonen, 18th September 2017). The Lutheran Church of Finland was in freefall and in the grip of the Left. In 2006, homosexuality wasn't really discussed in Finland. In Oulu, I frequently heard the view that 'gays go to Helsinki;' to the decadent, Godless south. Just four years later, the city saw its first

'North Pride' (titled in English rather than Finnish) and in 2018 this was attended by a group of Lutheran priests.

11. Juha Sipilä, Timo Soini, and the Great Betrayal

However, for many Finns, by 2015, the initial buzz of being enriched had died off and the reality had hit. Both the Centre Party and, of course, True Finns stood on a promise to reduce immigration. Yet, almost unbelievably, when a million so-called 'refugees' crossed the Mediterranean into the European Union, Sipilä went back on this solemn promise. Indeed, he went on national television to explain that it was Finland's duty to take a share of these (mostly) male economic migrants, with a share of Islamists among them who would go on to bomb a Paris rock concert that November. As I watched his broadcast, I couldn't quite believe what I was hearing. One had to have a certain degree of sympathy for the Prime Minister. His 20-year-old son had died during the election campaign only 5 months earlier. He was surely in the depths of grief. But that would be a reason to step down as Prime Minister and spend time with his family, not to perform a massive political U-turn against the clearly expressed will of the Finnish people.

Then, I thought, 'He'll have trouble doing what he wants to do anyway, because True Finns, an explicitly anti-immigration party, will not tolerate this. Their leader, Timo Soini (by then the Foreign Minister) will simply bring down the government.' To my amazement, he did no such thing. Instead, he permitted a policy change which we would expect to be anathema to almost everyone in his party. I met Timo Soini in January 2006 when he was campaigning for the Presidential Election on Rotuaari – Oulu's High Street - and I was working part-time for Oulu's English-language newspaper. At that time, an Englishman in Oulu – or anywhere in Finland – was still remarkable, and he seemed quite interested to talk to me. He spoke very good, if strongly accented, English to me in front of his (mainly working class or elderly) supporters and the supporters were clearly impressed that he could speak English at all, let alone so well. Soini was gregarious and seemed quite genuine. He explained how he'd spent time in the UK, where he began to support notoriously hooligan-ridden London football team Millwall (he even donned their scarf). Soini had also travelled around Ireland where he had undergone an intense religious experience and had become a

Roman Catholic, 'So, I know what it's like to be a minority,' he told me. I then travelled with him and one of his party workers, by car, to the Zeppelin shopping centre in Kempele, while he showed me the advert for him in *Ilta Sanomat,* which he thought looked rather good. 'Hei, Timo!' people called, enthusiastically, as I ambled round the mall with him. This was a 'man of the people.' Though a lawyer, he evidently understood the instincts of the average Finnish working class man.

It is beyond me why Soini did not bring down the government. Having spent his political career as a kind of populist-jester, albeit a just about socially acceptable one who most centrist Finns would never dream of terming 'far right,' Soini maybe realised that, in the hysteria of September 2015, such an act would make him beyond the political pale. Maybe, after all these years, the deep Finnish need to conform finally hit him, like a second conversion. But, as a consequence, in 2017, he was removed from the leadership of True Finns by its livid membership and replaced by the hardliner Jussi Halla-Aho. I met Hallo-Aho in 2011, during the General Election campaign that Spring. Cerebral, shy, ectomorphic and unwilling to speak English, he was a very different plate of *makkara* ('Finnish sausages') from Timo Soini. His difference from the, in essence, Establishment Soini was such that the other members of the government coalition were not prepared to have him as a cabinet member. The government almost fell, but it was saved by all the True Finns' government members, and about half their MPs, splitting to form a new party called 'Blue Reform,' a party which has minimal public support.

We, of course, know what happened as a consequence of these two betrayals. Preventable rapes, Oulu's grooming gangs and the almost successful attempt by the Finnish Establishment to cover up the suicide of a raped teenage girl . . . But, on the plus side, at least Oulu wasn't as boring as it had been in July 2003.

Chapter Seven

The Evolution of Rape

1. Why Has Finland Turned to Multiculturalism and Muslim Grooming?

From an evolutionary perspective, what has taken place in Oulu and how it has occurred and when it has occurred should be no surprise to anyone. As we have discussed, the specific evolutionary history of the Finns has meant – in essence – that they were under conditions of purifying Darwinian selection, and more intense conditions of this kind of selection, until more recently than is the case in a country such as the UK. As we have explored, this intense selection meant that Finns ended-up having high intelligence but, due to the need to be very strongly adapted to the ecology indeed, a small gene pool and thus very few high (or low) intelligence outliers. This severe yet predictable environment favoured highly cooperative groups, meaning low testosterone, high Agreeableness, high Conscientiousness, low Extraversion, and high Social Anxiety (resulting in high Neuroticism). This also led to relatively pronounced levels of schizophrenia (a sign of low testosterone and extreme empathy). This is because this 'negative' was outweighed by the 'positive' of the average Finn being very cooperative and high in Agreeableness and thus higher in schizotypy than the average person in a less harsh ecology. As such, Finns have been selected to be very low in the autism trait of low empathy. And, due to the small gene pool, there are relatively few personality outliers as well. This same ecology also favoured a relatively ethnocentric strategy, rather than one based around relatively low ethnocentrism but high levels of genius. All of this has meant a highly cooperative and conformist society with few per capita geniuses – these generally being outlier high IQ people with moderately low empathy and Conscientiousness. However, it also created a people capable of great self-sacrifice and even savagery to threatening outsiders.

As we have seen, the Industrial Revolution arrived in Finland late, meaning that it took longer for maladaptive inclinations, underpinned by spiteful mutations, to become prominent and for adaptive ideas, such as traditional religiousness and thus ethnocentrism, to begin to collapse. But, once they did, the lack of geniuses, the lack of difference in intelligence and personality and the massive levels of trust and social anxiety meant that Multiculturalism could spread extraordinarily quickly. And once it did, any critics of it – in such a group-think based society – would be rigorously suppressed, even to the point of the Finnish Establishment covering up the rape of children by Muslim immigrants and covering up the suicide of one of their rape victims. We can understand, therefore, why this process has occurred in Finland in the unusually rapid way that it has. But this begs a final question. Why do these Middle Eastern immigrants rape teenage girls at all?

2. Why Are There Muslim Grooming Scandals?

So far, we have dealt with understanding the nature of the Finns and why this nature has resulted in such a swift collapse into Multiculturalism. However, the Muslim Grooming of underage girls seems to eventually appear wherever there are large numbers of Muslim males in a European city. And, shockingly, research by psychologists and biologists implies that there might actually be sound evolutionary reasons why these Muslim grooming scandals have occurred, whether it is in Oulu or in Rotherham. This startling conclusion means that once Oulu permitted entry to large numbers of Muslim refugees then Muslim grooming was essentially inevitable. And it would be made even worse by the extremely trusting and conformist nature of the hosts.

Woodley of Menie et al.'s (2017) research would appear to imply that people from the Islamic world would have been under purifying Darwinian selection until even more recently than have Finns. This means, following their model, there would be relatively few of them with 'spiteful mutations' and we would expect them to act in a highly adaptive way. Moreover, data indicates that modern conditions, contraception plus a highly complex state, have effectively selected against fast Life History strategy and thus against high testosterone males. As we have discussed, an r-strategy invests his energy in copulation; he does not want to invest his energy in nurture. Prior to the development of a highly complex state, his investment in copulation would result in his passing on his genes, but nothing could compel him to invest in his offspring and he often wouldn't, forcing a single mother to fend for herself. This is now increasingly difficult because developed states can track down errant, r-strategist fathers and make them to financially invest in, and thus nurture, their offspring. Perhaps as a consequence, there is now a negative correlation in Western countries between fertility and markers of r-strategy among men: r-strategy men may have a lifetime of casual girlfriends but they will avoid having children. As a result, it will be the more feminine – *K*-strategy – males who will be more likely to pass on their genes (Woodley of Menie et al., 2017b). And it is among these increasingly feminized males that Arab refugees have come.

As an Arab refugee in Finland or any Western country, you are a male of extremely low social status, a point noted by psychiatrist Michaela Hynie (2018). According to evolutionary psychologist David Buss (1989) this makes you markedly unattractive to females, because, as we have seen, females are evolved to sexually select, to a much greater extent than males, for status. You can attempt to play for status – and, in my experience Arab refugees in Oulu do exactly this – by dressing as well as possible and 'strutting,' so indicating, on some level, that you are 'manly' and 'successful.' But, in general, your prospects of passing on your genes, according to Buss, are going to be very poor. As we have discussed, on the basis of your nationality alone, due to your origins in a low status society, many Finnish females will be repelled by you even if you do attain reasonable socioeconomic success. They would rather date a British computer programmer than an Iraqi one, even if the Iraqi one was better dressed and more handsome. Political scientist Kharam Iqbal (2015, p.111) has observed that in a society with a large number of single young men, youth gangs develop and they will often be violent, open to being drawn into terrorism under certain circumstances. Such gangs have been highlighted as a particular problem in India, for example (Patil, 2015). I noticed these in rural towns in India when I was there in July 2007: gangs of very well dressed young men, their brightly coloured shirts sparklingly clean, simply 'hanging around' with nothing to do. Such males, following the evolutionary models of David Buss, are going to be very unattractive to most females. Psychologist Lee Ellis (1989, p.53) in reviewing a large number of studies, concludes that, in this context, males will tend to adopt a strategy of passing on their genes through rape. In particular, this situation will favour gang rape, as this affords the rapist the protection of a gang, diluting the risk of being prosecuted, and it means that the victim can be more easily overpowered. Moreover, according to biologist Randy Thornhill and anthropologist Craig Palmer's literature review (Thornhill & Palmer, 2011, p.174) there is evidence that rape elevates the likelihood of a woman becoming pregnant, because the rapist produces more semen to compensate for the possibility that somebody else may have recently had sex with the victim. Ellis (1989, p.53) has conducted a detailed literature review which shows that being a low status single male is a key predictor of being a rapist and that predatory rapists – opportunists who do not know their victims well or at all – tend to be of particularly low status. This problem with such gangs has been argued to be one of the reasons why, in polygamous Islamic societies, women are subject to *purdah*. If they are covered up and chaperoned, then they won't be targeted by these gangs which develop because high status males will be able to monopolise the females (Shehabuddin, 2008, p.105). In the case of gang rape, studies of street gangs have found that the gang's most dominant male will rape the woman first with the remainder following in an approximate pecking order (Grewel, 2016). Ellis (1989) observes that assuming the female became pregnant, in a society in which abortion was illegal or expensive or dangerous, the gang members would all have some chance of passing on their genes this way. One could argue that this behaviour is useless in Finland, where the girl would most likely have an abortion if she became pregnant. However, these evolved behaviours are, according to Ellis, instinctive, and not subject to reasoned deliberations regarding outcomes. Ultimately, many of these young male refugees are in the circumstances where a propensity to gang rape makes clear adaptive sense. And then whom do you gang rape? According to David Buss (1989), men are evolved to be attracted to females who convey indications of fertility. Their chances of impregnating a woman are also elevated if she is a virgin.

The fact that these males are different ethnic group from the majority will even further elevate the probability of gang rape. According to a literature review by Frank Salter and Henry Harpending (2015), it is clear it is clear that people are disposed to act in order to elevate the interests of their ethnic group. If their ethnic group as a whole is of low status, then a key way in which they can elevate its status is through the gang rape of girls from the dominant ethnic group. Indeed, research by Congolese physician Denis Mukwege et al. (2010) implies that to divinely sanctify the rape of the invaded country's females is likely to be very adaptive in terms of group selection. Rape, they argue, is a way of asserting dominance not just over the females, but, by extension, over their fathers, brothers, male cousins and, in many ways, all males on the opposing side. It destroys their morale and undermines their confidence, because the conquerors assert dominance and control over the central resource for future existence, namely the wombs of the women of those whom they are conquering. Based on an analysis of the Democratic Republic of Congo, Mukwege et al. (2010) aver that rape can be a quite deliberate war strategy, because it creates deep trauma and insecurity among the victims and their networks, helping to undermine their ability to defend themselves. It may, therefore, be no coincidence that the original meaning of 'rape' was to 'pillage' or 'steal.' Only in the early 15th century did 'rape' come to refer to the abduction and sexual violation of a woman.

So, research by experts in evolutionary psychology predicts that if you put a large number of very low status males from the same ethnic group into a homogenous of a different ethnicity community then the result will be rape and, in particular, gang rape. Moreover, research by psychologist Daniel Nettle (2009) implies that it will be mainly working class girls who will be raped. According to Nettle, low socioeconomic status is predicted by a relative r-strategy and part of this strategy is sexual promiscuity, weak social bonds, early sexual maturation and being attracted to physical status – a man who is strong and can win fights in the unstable ecology you inhabit – rather than purely socioeconomic status. It follows that low status girls will be far more receptive, easier to access – their parents will be less concerned about their whereabouts, and, as they are less intelligent, such girls will be easier to manipulate. There will simply be less opposition to your activities, as starkly demonstrated in the Oulu and Rotherham examples. Psychologist Jonathan Haidt (2003) has observed that society at large is also less likely to care about such girls, because people of low social status evoke feelings of disgust. Elite and high socioeconomic status Finns express this disgust response with the derogatory terms which they use about their native working and under-class. They are *'juntti'* ('hicks' or 'yokals'), *pummi* ('bums'), *amis* (one who has attended a vocational high school rather than an academic one, the latter preparing you for higher education) and so on (see Dutton, 2009b). Many of the underage girls who were raped in Oulu were from impoverished, run down areas of the city, such as Tuira. So, based on examining the work of these experts, the conclusion must be that Oulu's Muslim Grooming Scandal was inevitable the moment significant numbers of young, single Muslim males were accommodated in the city.

3. Rape in the Name of God

Religiousness, as we have discussed, tends to sanctify as God's will evolutionarily adaptive behaviour. This, according to experts in the psychology of religion, appears to add another dimension to the inevitability of rape when a city accommodates large numbers of single males from an Islamic society. In a review entitled 'When Religion Makes It Worse,' Yael Sela and colleagues (Sela et al., 2015) observed that various parts of both the Old Testament, to which Muslims in part adhere or which have influenced Islam, and the Islamic scriptures, therefore, unsurprisingly render it God's will that you can rape the daughter of your enemy if you invade their land. It would surely elevate the damaged self-esteem of Muslim refugee males to believe that they are part of an invading force. It is indeed a common belief among fundamentalist Muslims that they must colonise the West under Islam (Armstrong, 2001). Leviticus 20: 13 tells believers that in a situation of war, you should kill every male in the opposing tribe and take all the females for yourself. Zechariah 14 is explicit that the enemies of Jerusalem are to be vanquished while their womenfolk are to be raped and enslaved. Judges 21 tells those who fear the Lord to invade the place of their enemies and kill every male as well as every female who is not a virgin. Virgins, however, are to be forcibly married to the soldiers who have slaughtered their families. The *Koran* 4:3 is quite clear that a man should take multiple wives: 'Marry of the women, who seem good to you, two or three or four; and if you fear that you cannot do justice (to so many) then one (only) or (the captives) that your right hands possess.' In other words, you can do what you like with female infidels, with the womenfolk in the country which you have invaded.

We have already discussed research from the Congo which explore the effectiveness of rape as a weapon of war. So, in light of research by psychologists such as Yael Sela and her colleagues, it makes sense that religion, as something highly adaptive, would

110

justify certain kinds of rape, specifically against outsiders. This would keep the in-group cooperative, but also allow low status males to procreate, making them less likely to rebel. In this regard, J. Philippe Rushton (2005) has averred that it would help to motivate single males to risk their lives in war for their group, as they would receive the female spoils and receive them with God's approval. In this sense, Rushton argues, it is surely no coincidence that many Islamic martyrs are single men (likely virgins, in a conservative society) and their reward in paradise is an extremely desirable harem and a permanent erection. This reward is explicit in the *Hadith,* in which the early Muslim community recorded the oral traditions of various things that the Prophet Mohammed had said when he was alive.

'Abu Umama narrated: "The Messenger of God said, 'Everyone that God admits into paradise will be married to 72 wives; two of them are houris and seventy of his inheritance of the [female] dwellers of hell. All of them will have libidinous sex organs and he will have an ever-erect penis.' (Sunan Ibn Majah, Zuhd, Book of Abstinence, 39).

'Anas said, Allah be well-pleased with him: The Messenger of Allah said, upon him blessings and peace: "The servant in Paradise shall be married with seventy wives." Someone said, "Messenger of Allah, can he bear it?" He said: "He will be given strength for a hundred." (Sifat al-Janna, al-`Uqayli in the Du`afa', and Musnad of Abu Bakr al-Bazzar).

'From Zayd ibn Arqam, Allah be well-pleased with him, when an incredulous Jew or Christian asked the Prophet, upon him blessings and peace, "Are you claiming that a man will eat and drink in Paradise?" He replied: "Yes, by the One in Whose hand is my soul, and each of them will be given the strength of a hundred men in his eating, drinking, coitus, and pleasure." (Sifat al-Janna, al-`Uqayli in the Du`afa', and Musnad of Abu Bakr al-Bazzar).

Horrific as Oulu's Muslim grooming scandal is, the evolutionary dimensions to it must not be forgotten. According to experts in this field, an influx of low status single men from a different race will surely elevate the likelihood of the grooming of underage girls and of gang rape. This will occur partly because rape is the only evolutionary strategy open to them, partly as a means of conquest, and partly because Islamic scriptures can be interpreted to justify it.

4. What Does the Future Hold?

The Finnish word for 'future' is *'tulevaisuus.'* I learnt this rather congenial-sounding word before I learnt the term for 'past.' In fact, I assumed that if 'future' was literally 'coming-vaisuus' then 'past' would surely be *'menevaisuus'* ('gone-vaisuus'). Naturally, I started using the word *menevaisuus* in semi-drunken conversation with people in bars; this being the main situation in which I speak Finnish. In actuality, the Finnish word for 'past' is *'menneisyys.'* But it was years before anyone corrected me on this. Finns were simply flattered, and probably slightly amused, that I was speaking their language at all. I speak in a pronounced English accent, my grammar is a train wreck, and I use English idioms in literal translation. But, as long as I'm speaking Finnish, they don't care, and they certainly won't be so impolite, as they would see it, to tell me that the word I am using for 'past' is wrong, and hilariously wrong at that.

So, what is the probable *tulevaisuus* for Oulu and for Finland as a whole? There are many studies which allow us to make reasonable predictions. And I'm afraid that in certain key respects it is a *tulevaisuus* which a lot of people are going to find troubling. In his book *Ethnic Conflicts,* Tatu Vanhanen (2012) has shown that the more ethnically diverse a society is, the higher is the degree of ethnic conflict. Indeed, the correlation between a country's ethnic diversity and its level and intensity of ethnic conflict is 0.66. The more ethnically diverse Finland becomes, the more conflict-ridden it will be and the more these conflicts will be based around ethnicity. In such circumstances, in which two groups are in conflict, Vanhanen shows that people tend to identify more strongly with their ethnic group and are more likely to perceive outsiders as an enemy. Thus, his research indicates that as the population of Finland becomes more ethnically-diverse, many Finns will probably develop their sense of Finnishness and become more nationalistic, while the foreigners will be decreasingly likely to integrate and will feel decreasingly Finnish. This is a recipe for a spiral into increasingly intense ethnic conflict; into low-level civil war.

But ethnic diversity also has an effect on other aspects of how societies operate. Putnam (2007) has shown that ethnic diversity dramatically reduces trust and makes people much less inclined to invest in the 'common good.' The result of this is that 'Welfare States' tend to collapse. Welfare States are predicated upon high levels of trust. People are prepared to pay tax so that others, whom they trust and see as members of their in-group, can be assisted in hard times. They do this because they trust that these people will do the same for them. To some extent, a welfare state involves looking after your own genetic interests. By funding a health service, for example, you are assisting your ethnic kin. But once you are no longer doing this in such a clear cut way, because the society is ethnically diverse, then the incentive to contribute to the common good is reduced (Putnam, 2007). But, it gets worse because, as we have discussed and as Putnam (2007) shows, ethnic diversity reduces trust even among the natives. It creates a situation where every native is less trustworthy because every native is a potential defector. Moreover, there is an extent to which people blame other natives for having let this problematic situation of ethnic diversity occur at all. This means that generalized trust is reduced. This will also result in the gradual erosion of the Welfare State until there isn't really much of a Welfare State at all.

As this happens, a further factor will likely introduce itself to force down levels of trust. Eric Ulsuner (2000) has shown, as we have already noted, that societies higher in socioeconomic equality are also higher in trust. But as the Welfare State declines, there will inevitably be greater socioeconomic inequality. Currently, in Finland, if you put your child in daycare you pay fees but these vary according to your perceived ability to pay, based on what you earn. As the welfare state collapses, so will policies of this kind. Everybody will simply pay the same and, over time, very substantial differences in living standards will start to become evident, as will social resentment and even resentment of wealthier areas of the country by poorer areas of the country. This social resentment is also likely to lead to an increase in crime.

This would happen in any time period, but in an electronic age these processes are likely to occur much faster and have more profound effects. As trust levels continue to plummet, people stop trusting mainstream politicians and mainstream newspapers, a process we are already observing. They will begin to assume – as a matter of course – that these people and news sources are deceiving them and, as we have seen, on many key matters they will be correct. They will turn to alternative news sources and vote for alternative political figures. As jurist Cass Sunstein (2018) has explored, this will create an echo chamber effect whereby people will increasingly only be hearing the viewpoints which they already accept. This will help to further cement a divided, Balkanized, and untrusting society. Concomitantly, we will expect Finnish society – like all European societies – to become increasingly diverse in terms of worldview, as the 'spiteful mutants' spread their views which, through a virtue-signalling arms race to appear ever more caring, will become more and more extreme. This follows 'O'Sullivan's First Law,' that any institution that is not explicitly right wing will always drift leftwards over time (Will, 1994, p.280). In the year 2000, gay marriage was pretty much unthinkable. By 2010, one could expect to be heavily publically criticised for opposing it, as happened to Päivi Räsänen. However, in 2010 many now accepted ideas about transgenderism were unthinkable. What is unthinkable now is likely to be dogma a decade hence. This process will elevate political divergence, distrust and division.

Indeed, trust is central to democracy and these processes have already damaged trust considerably, and will continue to disenfranchise citizens to the point that they may increasingly lose faith in democracy, regarding it as corrupt and a waste of time. They will vote for political parties that are suspicious of democracy, at least when the country is ruled by those whom they perceive to be enemies. Depending on their perspective, these parties will be of the far right or far left by current standards. Movements not unlike the Lapua Movement will gain traction in circumstances like this, as will groups from the extreme left. And the kind of generally wealthy people who become involved in Finnish politics – in Finland individual candidates must fund their own campaigns – will be increasingly distrusted and perceived as, essentially, 'other;' as not part of the in-group. This will, unfortunately, have ramifications for their safety, especially as people begin to blame those who advocated and enforced Multiculturalism for its future consequences.

This distrust will start to spread, like a disease, to other organs of the Finnish state. The police will be increasingly perceived – especially if they are seen to enforce unpopular policies with regard to not being allowed to criticise immigration – not as protectors of the Finnish people but as 'enemies of the people.' Distrust of the police, will result in the establishment of para-military forces of the kind that have been so prevalent in the recent history of Northern Ireland. This can already be observed with patrols in Finland by the Soldiers of Odin. If this seems too far fetched, it should be remembered that in the 1970s, trust in the police in strongly Catholic parts of Northern Ireland was so low that the British state simply had no control in many of these areas. Former British Chancellor of the Exchequer Ken Clarke (2016, p.56) recalls of a visit to Londonderry in 1972 that, 'large parts of Derry were effectively under the control of the IRA and were officially designated no-go areas for both the police and the army.' In a society with very low trust, this is the kind of thing that happens.

It is extremely stressful to live in a conflict-ridden, highly unequal society in which there are very low levels of trust. As we have already seen, stress – and also feelings of exclusion – are the key environmental factors behind religiousness (Norenzayan & Shariff, 2008). As such, for a portion of Finnish society, some form of genuine religious belief would be elevated and this would, in turn, increase both positive and negative ethnocentrism. At the same time, the correlation between religiousness, which is 40% genetic, and fertility is likely to mean that Finns will become increasingly religious simply for genetic reasons as well. Moreover, if we are stressed we tend to be more prone to our cognitive biases, including religiousness. One such cognitive bias, and as we have discussed why there is reason to think that it is higher among the Finns than it is among European ethnic groups further south, is ethnocentrism. So, as stress is elevated, this may start to *increase* among a portion of the population.

Meanwhile, the negative correlation between fertility and intelligence will mean that the Finns of the future will, anyway, be less trusting, less open-minded, and more negatively ethnocentric. This is because low intelligence predicts negative ethnocentrism (Hodson & Busseri, 2012), possibly because low intelligence predicts low trust in general. However, low IQ also predicts support for nationalist political parties (Deary et al., 2008). So it may be that such people are simply more governed by evolved cognitive biases due to the way in which intelligence, and problem solving, involves rising above such biases (Dutton & Van der Linden, 2017). Computer models have demonstrated that once 25% of a group adhere to a counter-cultural viewpoint, such as ethnocentrism currently is, and become 'activists' in fervently advocating it, these activists gradually tip the opinion of the entire group towards their own (Centola et al., 2018). At this critical mass, they can disrupt the transmission of and faith in the majority view to such an extent that more and more people begin to change sides, tipping the opinion of the majority to that of the minority. Philosopher Sean Gabb (2007) has observed how this method of the 'March Through the Institutions' was articulated by the Italian Marxist Antonio Gramsci (1891-1937). Gabb observes that such tactics could now be used to undo the dominance of Multiculturalism. A more stressful and religious society is likely to ensure that this critical mass of ethnocentrics is reached. In this way, influential institutions can be gradually taken over from beneath. And as we have observed (Hammond & Axelrod, 2006), based on computer models, all else controlled for, the more ethnocentric group will always triumph. As discussed, the Finns are high in conformism and so when the country, or a significant part of it, 'flips' from Multiculturalism back to nationalism, it will happen very quickly.

5. 'So Sadly Neglected . . .'

The first time I was interviewed for a Finnish national newspaper was in September 2008 after I had written a piece for *The Guardian* (Dutton, 23rd September 2008) on the cultural causes of a school massacre that had just happened in Finland. The *Ilta-Sanomat* journalist who interviewed me asked me, 'What should we do?' I was taken aback by the whole incident. Not yet realising just how concerned Finns are about how others perceive them, it had never occurred to me that a Finnish journalist would even read my article, let alone want to write a report on it. And I was particularly struck that he should be interested in my opinion – the outsider's opinion – on what *his* people should do about the country's second school shooting in a year. I told him that it wasn't up to me, as a foreigner, to tell Finns what to do. But there was a degree to which, as an outsider, I could maybe see what was going on more clearly than Finns could, because I didn't suffer from 'home blindness.' This is the phenomenon whereby if you've been raised in a culture you are less able to perceive it objectively and you are thus less able to objectively analyse and make sense of it (Kapferer, 2001). As far as I

can see, the simplest explanation for the speed with which Oulu –
and Finland – changed from nationalist to Multiculturalist is the
specific national psychology of Finland, moulded under intense
conditions of Darwinian selection. And it is possible that this same
national psychology – including a relatively ethnocentric strategy
and a high degree of conformism when the dominant ideology
begins to change - will result in a new National Awakening also
coming more quickly than many might predict.

References

Agarwal, S. (2013). High prevalence of low testosterone levels in male patients with schizophrenia. *European Psychiatry,* 23: Supplement 1.

Ålands Nyheter. (5th December 2018). Finnish prime minister condemns child sex abuse in city of Oulu. https://www.alandsnyheter.com/in-english/finnish-prime-minister-condems-child-sex-abuse-in-city-of-oulo/

Ålands Nyheter. (4th December 2018). Eight enemy soldiers arrested for raping Finnish schoolgirl in Oulu. https://www.alandsnyheter.com/in-english/eight-enemy-soldiers-arrested-for-raping-finnish-schoolgirl-in-oulu/

Alestalo, M. & Toivonen, T. (1977). *Changes in Class Structure and Stratification in Finnish Society 1950-1974.* Helsinki University: Department of Sociology.

Almlund, M., Duckworth, A., Heckman, J. & Kautz, T. (2011). Personality, psychology and economics. In Hanushek, S., Machin, S. & Woesmann, L. (Eds.). *Handbook of the Economics of Education,* Amsterdam: Elsevier.

Annala, P. (3rd December 2018). Poliisilla on lisää epäiltyjä alle 15-vuotiaan tytön törkeästä raiskauksesta Oulussa – seitsemän yhä vangittuna. *YLE,* https://yle.fi/uutiset/3-10536727

Anttonen, P. (2005). *Tradition through Modernity: Postmodernism and the Nation State in Folklore Scholarship.* Helsinki: Finnish Literature Society.

Armstrong, K. (2001). *The Battle for God: Fundamentalism in Judaism, Christianity and Islam.* London: HarperCollins.

Aro, T. (1985). The race of the Finns in German and Nordic encyclopaedias. Kemiläinen, A. (Ed.). *Mongols or Germanics?* Helsinki: Finnish Historical Society.

Austin, D.F.C. (1996). *Finland as a Gateway to Russia: Issues in European Security.* Avebury.

Bachmann, S. Cross, C., Kalbassi, S., Sarraf, M., Woodley of Menie, M.A. & Baudouin, J.M. (2018). Protein pheromone MUP20/Darcin is a vector and target of indirect genetic effects in mice. *Preprints,* https://www.biorxiv.org/content/10.1101/265769v1

Badcock, C. (2003). Mentalism and mechanism: Twin modes of human cognition. In Crawford, C. & Salman, C. (Eds.).

Human nature and social values: Implications of evolutionary psychology for public policy. Mahwah, NJ: Erlbaum.

Barnett, A. (2007). *Sibelius.* New Haven, CT: Yale University Press.

Baron-Cohen, S. (2002). The extreme male brain theory of autism. *Trends in Cognitive Sciences,* 6: 248-254.

Berggren, N., Jordahl, H. & Poutvaara, P. (2017). The Right Look: Conservative Politicians Look Better and Voters Reward It. *Journal of Public Economics*, 146: 79-86.

Bergmann, U. (2016). *Nordic Nationalism and Right-Wing Populist Politics: Imperial Relationships and National Sentiments.* London: Palgrave Macmillan.

Blanchard, R. (2008). Review and theory of handedness, birth order, and homosexuality in men. *Laterality,* 13: 51-70.

Boyer, P. (2001). *Religion Explained: The Human Instincts that Fashion Gods, Spirits and Ancestors.* London: William Heinemann.

Broman, S., Brody, N., Nichols, P. et al. (1987). *Retardation in Young Children.* Hillsdale, NJ: Erlbaum.

Bruhn, S. (2003). *Saints in the Limelight: Representations of the Religious Quest on the Post-1945 Operatic Stage.* Hillsdale, NY: Pendragon Press.

Bullock, D. (2014). *The Russian Civil War, 1918-1922.* London: Bloomsbury.

Buss, D. (1989). *The Evolution of Desire: Strategies of Human Mating.* New York: Basic Books.

Cabeza de Baca, T. & Woodley of Menie, M.A. (2017). WITHDRAWN: Life history as the appropriate meta-theory for evaluating group-differences in psychopathic personality: A comment on Lynn (2017). *Journal of Criminal Justice,* https://doi.org/10.1016/j.jcrimjus.2017.05.007

Calhoun, J. (1973). Death Squared: The explosive growth and demise of a mouse population. *Proceedings of the Royal Society of Medicine,* 66: 80-88.

Centola, D., Becker, J., Brackbill, D. & Baronchelli, A. (2018). Experimental evidence for tipping points in social convention. *Science,* 360: 1116-1119.

Cheon, B., Livingston, R., Hong, Y. & Chiao, J. (2014). Gene × environment interaction on intergroup bias: the role of 5-HTTLPR and perceived outgroup threat. *Social Cognitive and Affective Neuroscience,* 9: 1268-1275.

Chiao, J. & Blizinsky, K. (2010). Culture–gene coevolution of individualism–collectivism and the serotonin transporter gene. *Proceedings of the Royal Society, B.* DOI: 10.1098/rspb.2009.1650

China.org. (2nd October 2015). Refugee protests lead to revised food menus in Finland. http://www.china.org.cn/world/Off_the_Wire/2015-10/02/content_36729327.htm

Church Times. (27th October 2010). Thousands resign from Church of Finland.

Clark, G. (2014). *The Son Also Rises: Surnames and the History of Social Mobility.* Princeton, NJ: Princeton University Press.

Clarke, K. (2016). *Kind of Blue: A Political Memoir.* London: MacMillan.

Clements, J. (2014). *An Armchair Traveller's History of Finland.* London: Haus.

Cochran, G. & Harpending, H. (2009). *The 10,000 Year Explosion: How Civilization Accelerated Human Evolution.* New York: Basic Books.

Crespi, B. (2016). Autism as a disorder of high intelligence. *Frontiers in Neuroscience,* https://doi.org/10.3389/fnins.2016.00300

Cukic, I. & Bates, T. (2015). The association between Neuroticism and heart rate variability is not fully explained by cardiovascular disease and depression. *PLoS One,* 10(5): e0125882.

Danver, S. (Ed.). (2015). Finland. In Danver, S. (Ed.). *Native Peoples of the World: An Encylopedia of Groups, Cultures and Contemporary Issues: An Encylopedia of Groups, Cultures and Contemporary Issues.* London: Routledge.

De Fruyt, F. & Mervielde, I. (1996). Personality and interests as predictors of educational streaming and achievement. *European Journal of Personality,* 10: 405-425.

Dearden, L. (17th July 2017). Home Office had information of Rotherham grooming gang in 2002 but failed to act, review finds. *Independent,* https://www.independent.co.uk/news/uk/crime/rotherham-grooming-gangs-inquiry-home-officer-information-2002-failure-a8451926.html

Deary, I., Batty, G. & Gales, C. (2008). Childhood intelligence predicts voter turnout, voter preferences and political involvement in adulthood; the 1970 cohort. *Intelligence* 36: 548-555.

Dodds, B. (2008). Patterns of Decline: Arable Production in England, France and Castile, 1370–1450. In Dodds, B. & Britnell, R. (Eds). *Agriculture and Rural Society After the Black Death: Common Themes and Regional Variations.* Hatfield: University of Hertfordshire Press.

Dowson, J. & Grounds, A. (2006). *Personality Disorders: Recognition and Clinical Management.* Cambridge: Cambridge University Press.

Dunkel, C. & Dutton, E. (2016). Religiosity as a predictor of ingroup favoritism within and between religious groups. *Personality and Individual Differences,* 98: 311-314.

Dutton, E. (In Press). *Race Differences in Ethnocentrism.* Budapest: Arktos Publishing.

Dutton, E. (2018). Fellatio. In Shackelford, T. & Shackelford-Weekes, V. (Eds.). *Encyclopaedia of Evolutionary Psychological Science.* New York: Springer.

Dutton, E. (2015a). Obituary: Tatu Vanhanen (1929-2015). *Mankind Quarterly,* 56: 225-232.

Dutton, E. (2015b). *The Ruler of Cheshire: Sir Piers Dutton, Tudor Gangland and the Violent Politics of the Palatine.* Northwich: Léonie Press.

Dutton, E. (5th December 2014). Finns Quit Church Over gay Marriage.

Dutton, E. (2012). Recent Ethnographic Research and the Modern Finnish Social Class System as an Evolutionary Strategy. *Acta Ethnographica Hungarica,* 57: 139-156.

Dutton, E. (October 2009). Playing the Blame Game: Finland the Soviets. *History Today,* 59: 10.

Dutton, E. (2009a). *Four Immigrant Churches and a Mosque: An Overview of Immigrant Religious Institutions in Oulu.* Turku: Finnish Institute of Migration.

Dutton, E. (2009b). *The Finnuit: Finnish Culture and the Religion of Uniqueness.* Budapest: Akademiai Kiado.

Dutton, E. (23rd September 2008). Violent male culture may be at root of Finnish school massacre. *The Guardian,* https://www.theguardian.com/education/2008/sep/23/finland.school.shooting.comment

Dutton, E. (26th March 2008). Friday Prayers at Oulu's Mosque. *65 Degrees North,* www.65degreesnorth.com (No longer online).

Dutton, E. (2008). Battling to be European: Myth and the Finnish Race Debate. *Antrocom: Online Journal of Anthropology,* 5: 1, http://www.antrocom.net/upload/sub/antrocom/040208/14-Antrocom.pdf

Dutton, E., Madison, G. & Van der Linden, D. (In Press). Why do high IQ societies differ in intellectual achievement? The role of schizophrenia and left-handedness in per capita scientific publications and Nobel prizes. *Journal of Creative Behavior.*

Dutton, E., Shibaev, V. & Becker, D. (2019). Why do Finns, Estonians and Finno-Ugric Peoples in Russia Have Such High Intelligence? Unpublished Manuscript.

Dutton, E. & Woodley of Menie, M.A. (2018). *At Our Wits' End: Why We're Becoming Less Intelligent and What It Means For the Future.* Exeter: Imprint Academic.

Dutton, E. & Madison, G. (2018a). Why do middle-class couples of European descent adopt children from Africa and Asia? Some Support for the Differential *K* Model. *Personality and Individual Differences,* 130: 156-160.

Dutton, E. & Madison, G. (2018b). Execution, Violent Punishment and Selection for Religiousness in Medieval England. *Evolutionary Psychological Science,* 4: 83-89.

Dutton, E., Madison, G. & Dunkel, C. (2018). The Mutant Says in His Heart, "There Is No God": The Rejection of Collective Religiosity Centred Around the Worship of Moral Gods is Associated with High Mutational Load. *Evolutionary Psychological Science,* 4: 233-244.

Dutton, E. & Madison, G. (2017). Why do Finnish men marry Thai women but Finnish women marry British men? Cross-National Marriages in a Modern Industrialized society exhibit sex-dimorphic sexual selection according to primordial selection pressures. *Evolutionary Psychological Science,* 3: 1-9.

Dutton, E. & Van der Linden, D. (2017). Why is Intelligence Negatively Associated with Religiousness? *Evolutionary Psychological Science,* 3: 392-403.

Dutton, E., Madison, G. & Dunkel, C. (2018b). The Mutant Says in His Heart, "There Is No God": The Rejection of Collective

Religiosity Centred Around the Worship of Moral Gods is Associated with High Mutational Load. *Evolutionary Psychological Science,* 4: 233-244.

Dutton, E., Madison, G. & Lynn, R. (2016a). Demographic, economic, and genetic factors related to national differences in ethnocentric attitudes. *Personality and Individual Differences,* 101: 137-143.

Dutton, E., Van der Linden, D., Madison, G., Antfolk, J. & Woodley of Menie, M.A. (2016b). The intelligence and personality of Finland's Swedish-speaking minority. *Personality and Individual Differences,* 97: 45-49.

Dutton, E. & Charlton, B. (2015). *The Genius Famine: Why We Need Geniuses, Why They're Dying Out and Why We Must Rescue Them.* Buckingham: University of Buckingham Press.

Dutton, E. & Lynn, R. (2014). Regional Differences in Intelligence and their Social and Economic Correlates in Finland. *Mankind Quarterly,* 54: 3.

Dutton, E., te Nijenhuis, J. & Roivainen, E. (2014). Solving the puzzle of why Finns have the highest IQ, but one of the lowest number of Nobel prizes in Europe. *Intelligence,* 46: 192-202.

Dutton, E. & Lynn, R. (2013). A Negative Flynn Effect in Finland, 1997-2009. *Intelligence,* 41: 5.

Ekelund, J., Lichterman, D., Hovatta, I., Ellonen, P., Suvisaari, J., Terwilliger, J., et al. (2000). Genome-wide scan for schizophrenia in the Finnish population: evidence for a locus on chromosome 7q22. *Human Molecular Genetics,* 7: 1049-1057.

Ellis, L. (1989). *Theories of Rape: Inquiries Into the Causes of Sexual Aggression.* New York: Hemisphere Publishing.

EVL. (28th January 2019). Kirkkoon kuuluu 69,7 prosenttia suomalaisista. https://evl.fi/uutishuone/tiedotearkisto/-/items/item/25401/Kirkkoon+kuuluu+69-7+prosenttia+suomalaisista

Fernandes, H., Lynn, R. & Hertler, S. (2018). Race differences in anxiety disorders, worry, and social anxiety: An examination of the Differential-*K* Theory in Clinical Psychology. *Mankind Quarterly,* 58: 466-500.

Finland Times. (1st October 2015). Refugees stage demonstration in Oulu.

http://www.finlandtimes.fi/national/2015/10/01/20908/Refuge
es-stage-demonstration-in-Oulu

Freeborn, D. (1998). *From Old English to Standard English: A Course Book in Language Variation Across Time.* Ottawa: University of Ottawa Press.

Friedman, H., Tucker, J., Tomlinson-Keasey, C. et al. (1993). Does childhood personality predict longevity? *Journal of Personality and Social Psychology*, 65: 176-185.

Gabb, S. (2007). *Culture Revolution, Culture War: How the Conservatives Lost England and How to Get It Back Again.* London: Hampden Press.

Ganley, C., Mingle, L., Ryan, A. et al. (2013). An examination of stereotype threat effects on girls' mathematics performance. *Developmental Psychology*, doi: 10.1037/a0031412

Gardner, H. (1983). *Frames of Mind: The Theory of Multiple Intelligences.* New York: Basic Books.

Gebauer, J., Bleidorn, W., Gosling, S. et al. (2014). Cross-Cultural variations in Big Five relationships with religiosity: A sociocultural motives perspective. *Journal of Personality and Social Psychology,* 107: 1064-1091.

Gleason, H.A. (1969). *An Introduction to Descriptive Linguistics.* London: Hodder & Stoughton.

Gottfredson, L. (1997). Editorial: Mainstream science on intelligence. *Intelligence*, 24: 13-24.

Grasten, H. (4th December 2018). Sisäministeri Mykkänen haluaa tarkemman seulan kansalaisuuden myöntämiseen - Oulun raiskaukset järkyttävät: "Tällaista ei pitäisi tapahtua Suomessa." *Ilta-Lehti,* https://www.iltalehti.fi/politiikka/a/f340531e-792b-4839-8408-f46bd01188e1

Grewel, K. (2016). *Racialised Gang Rape and the Reinforcement of Dominant Order: Discourses of Gender, Race and Nation.* London: Routledge.

Gugliemino, C.R. (1990). Uralic Genes in Europe. *American Journal of Physical Anthropology,* 83:1.

Gullberg, T. (2011). The Holocaust as History Culture in Finland. In Bjerg, H., Lenz, C. & Thorstensen, E. (Eds.). *Historicizing the Uses of the Past: Scandinavian Perspectives on History Culture, Historical Consciousness and Didactics of History Related to World War II.* Transcript: Time, Meaning, Culture.

Haarman, H. (2016). *Modern Finland*. Jefferson, NC: McFarland Publishing.

Haidt, J. (2003). The moral emotions. In Davidson, R., Scherer, K. & Goldsmith, H. (Eds.). *Handbook of affective sciences*. Oxford University Press.

Hammond, R. & Axelrod, R. (2006). The evolution of ethnocentric behaviour. *Journal of Conflict Resolution,* 50: 1-11.

Harmon, L. R. (1961). The High School Background of Science Doctorates: A Survey Reveals the Influence of Class Size, Region of Origin, as Well as Ability, in PhD Production. *Science* 133: 679–688.

Helmreich, W. (1982) *The Things They Say Behind Your Back: Stereotypes and the Myths Behind Them.* New York: Doubleday.

Hentilä, S. (2001). Finland's leap into democracy. *Virtual Finland,* http://www.finland.fi/netcomm/news/showarticle.asp?intNWS AID=25908

Helsingin Sanomat. (11th May 2011). Police find no evidence of crime in comments by True Finns MP Hakkarainen.

Helsingin Sanomat. (2nd May 2006). No prosecution for Finnish group for publication of Mohammed caricatures (English edition).

Helsingin Sanomat. (16th February 2006). Vanhanen apology over Muhammed cartoons was PM's own decision (English edition).

Helsingin Sanomat. (12th August 2004). Comments in interview could bring charges of inciting racism against PM Vanhanen's father (English edition).

Hietala, H. (11th May 2015). Oulun kirvessurmat: Kaaos pubissa alkoi yhtäkkiä ja päättyi minuuteissa – näin tapahtumat etenivät. *Ilta Sanomat,* https://www.is.fi/kotimaa/art-2000000923916.html

Hietala, M. (1985). In search on the Finnish female type. Kemiläinen, A. (Ed.). *Mongols or Germanics?* Helsinki: Finnish Historical Society.

Hills, P., Francis, L., Argyle, M. & Jackson, C. (2004). Primary personality trait correlates of religious practice and orientation. *Personality and Individual Differences,* 36: 61-73.

Hodge, M. (6th December 2018). VILE ATTACK: Girl, 10, repeatedly raped by 'migrant grooming gang' in Finland as cops warn parents to watch out for 'foreign men' contacting underage kids. *The Sun,* https://www.thesun.co.uk/news/7917239/migrant-grooming-gang-oulu-finland/

Hodgekins, J. (2015). Schizotypy and psychopathology. In Mason, O. & Claridge, G. (Eds.). *Schizoptypy: New Dimensions.* London: Routledge.

Hodson, G. & Busseri, M. (2012). Bright minds and dark attitudes: Lower cognitive ability predicts greater prejudice through right-wing ideology and low intergroup contact. *Psychological Science,* 23: 187-195.

Honkanen, H. (15th January 2019). Katie Hopkins and Local Patriots Expose Muslim Grooming Gangs in Finland. *VDare,* https://vdare.com/articles/katie-hopkins-and-local-patriots-expose-muslim-grooming-gangs-in-finland

Huntford, R. (1971). *The New Totalitarians.* New York: Stein & Day.

Hynie. M. (2018). The Social Determinants of Refugee Mental Health in the Post-Migration Context: A Critical Review. *Canadian Journal of Psychiatry,* 63: 297-303.

Ilta-Lehti. (6th April 2008). Oulun raiskaukset yhä selvittämättä. https://www.iltalehti.fi/oulu/a/200804067481452

Insall, T. & Salmon, P. (2011). *The Nordic Countries in the Early Cold War, 1944-51.* London: Routledge.

Iqbal, K. (2015). *The Making of Pakistani Human Bombs.* Lanham, MA: Lexington Books.

Jakobson, M. (1987). *Finland: Myth and Reality.* Helsinki: Otava.

Jamal, A. & Noorudin, A. (2010). The Democratic Utility of Trust: A Cross-National Analysis. *Journal of Politics, 72: 45-59.*

James, W. (1988). Testosterone levels, handedness and sex ratio at birth. *Journal of Theoretical Biology,* 133: 261-266.

Jensen, A. R. (1998). *The g Factor: The Science of Mental Ability.* Westport, CT: Praeger.

Jones, M. (1977). *Finland: Daughter of the Sea.* London: Dawson - Archon Books.

Julku, M. (4th December 2018). Tutkinnanjohtaja *IL:*lle: Osa Oulun seksuaalirikoksista epäillyistä oli jo ehtinyt saada Suomen kansalaisuuden - "Poikkeuksellinen monella tavalla." *Ilta-Lehti,* https://www.iltalehti.fi/kotimaa/a/001c56ca-c743-4e4b-9c1f-4f01ebd103a9

Jussilä, O., Hentilä, S. & Nevakivi, J. (1999). *From Grand Duchy to a Modern State: A Political History of Finland Since 1809.* London: C. Hurst & Co.

Jussim, L. (2012). *Social Perception and Social Reality: Why Accuracy Dominates Bias and Self-Fulfilling Prophecy.* Oxford: Oxford University Press.

Kaleva. (26th December 2018). Oulun kirkkoherrat osoittavat tukensa muslimeille moskeijan ilkivallassa – "Tässä yhteiskunnassa ihmisillä on oikeus harjoittaa uskontoaan vapaasti". https://www.kaleva.fi/uutiset/oulu/oulun-kirkkoherrat-osoittavat-tukensa-muslimeille-moskeijan-ilkivallassa-tassa-yhteiskunnassa-ihmisilla-on-oikeus-harjoittaa-uskontoaan-vapaasti/812624/

Kanazawa, S. (2012). *The Intelligence Paradox: Why the Intelligent Choice Isn't Always the Smart One.* Hoboken, NJ: Wiley & Sons.

Kannisto, P. & Kannisto, S. (2012). *Free as a Global Nomad: An Old Tradition with a Modern Twist.* Pheonix, AZ: Drifting Sands Press.

Kapferer, B. (2001). Star Wars: About Anthropology, Culture and Globalization. *Suomen Antropologi: Journal of the Finnish Anthropological Society,* 26: 2-29.

Karpinski, R., Kolb, A., Tetreault, N. & Borowski, T. (2018). High intelligence: A risk factor for psychological and physiological overexcitabilities. *Intelligence,* 66: 8-23.

Kaufman, S., DeYoung, C., Reiss, D. & Gray, J. (2011). General intelligence predicts reasoning ability for evolutionarily familiar content. *Intelligence,* 39: 311-322.

Kauranen, A. (25th September 2015). 'Finland's no good': Disappointed migrants turn back. *AFP,* https://news.yahoo.com/finlands-no-good-disappointed-migrants-turn-back-152042061.html?guccounter=1

Kääriäinen, H., Muilu, J., Perola, M. & Kristiansson, K. (2017). Genetics in an isolated population like Finland: a different basis for genomic medicine? *Journal of Community Genetics,* 8: 319-326.

Kemiläinen, A. (2000). *Finns in the Shadow of the 'Aryans': Race Theories and Racism.* Helsinki: Finnish Literature Society

Kent, N. (2014). *The Sámi Peoples of the North: A Social and Cultural History.* London: C. Hurst & Co.

Kerminen, S., Hauvulinna, A., Hellenthal, G., Martin, A., Sarin A.-P. et al. (2017). Fine-Scale Genetic Structure in Finland. *Genes, Genomes, Genetics*, 7: 3459–3468.

Kilpeläinen, J. (1985). Racial theories about the Western Finno-Ugric peoples in Central European anthropology in the 19th century and Finnish reactions to them. In Kemiläinen, A. (Ed.). *Mongols or Germanics?* Helsinki: Finnish Historical Society.

Kirby, D. (2006). *A Concise History of Finland.* Cambridge: Cambridge University Press.

Kirby, D. (1980). *Finland in the Twentieth Century: A History and an Interpretation.* Minneapolis: University of Minnesota Press.

Kittles, R, Perola, M., Peltonen, L. & Bergen, A. (1998). Dual Origins of Finns Revealed by Y-Chromosome Haplotype Variation. *American Journal of Human Genetics,* 62: 1171-1179.

Kivistö, K. & Makelä, K. (1967). Pappi lukkari talonpoika kuppari. *Sosiologia,* 4: 133-135.

Korpela, J. (2018). *Slaves from the North: Finns and Karelians in the East European Slave Trade, 900–1600.* Leiden: BRILL.

Korpela, J. (2014). The Baltic Finnic People in the Medieval and Pre-Modern Eastern European Slave Trade. *Russian History* 41: 1.

Kuisma, M. (2013). "Good" and "Bad" Immigrants: The Economic Nationalism of the True Finns' Immigration Discourse. In U. Korkut, G. Bucken-Knapp, A. McGarry, J. Hinnfors & H. Drake. (Eds.). *The Discourses and Politics of Migration in Europe.* London. Palgrave Macmillan.

Kukkonen, W. (1969). *Postwar Finland as presented in The New York Times and The Times, London, and a Comparison of Editorial Analyses.* Madison: University of Wisconsin Press.

Kura, K., te Nijenhuis, J. & Dutton, E. (2019). Spearman's hypothesis tested comparing 47 regions of Japan using a sample of 14 million children. *Psych,* 1: 26-34.

Kura, K., te Nijenhuis, J. & Dutton, E. (2015). Why do Northeast Asians Win So Few Nobel Prizes? *Comprehensive Psychology,* 4. http://www.amsciepub.com/doi/pdf/10.2466/04.17.CP.4.15

Laine, T. (2006). 'Shame on us': Shame, National Identity and the Finnish Doping Scandal. *International Journal of the History of Sport,* 23:1.

Lamnidis, T., Majander, K., Jeong, C., Salmela, E., Wessman, A. et al. (2018). Ancient Fennoscandian genomes reveal origin and spread of Siberian ancestry in Europe. *Nature Communications,* 9: 5018.

Larsen, D.L. (1998). *The Company of Preachers. Volume II.* Grand Rapids, MI: Kregel Publications.

Lavery, J. (2006). *The History of Finland.* Westport, CT: Greenwood Publishing.

Laythe, B., Finkel, D. & Kirkpatrick, L. (2001). Predicting prejudice from religious fundamentalism and right wing authoritarianism: A multiple regression analysis. *Journal for the Scientific Study of Religion,* 40: 1-10.

Leinonen, A. (1st December 2018). Poliisi tutkii lapseen kohdistunutta törkeiden seksuaalirikosten sarjaa Oulussa – seitsemän miestä on vangittu. *Kaleva,* https://www.kaleva.fi/uutiset/oulu/poliisi-tutkii-lapseen-kohdistunutta-torkeiden-seksuaalirikosten-sarjaa-oulussa-seitseman-miesta-on-vangittu/811321/

Levin, M. (2005). *Why Race Matters: Race Differences and What They Mean.* Oakland, VA: New Century Foundation.

Lukaszewski, A., Gurven, M. Von Rueden, C. & Schmitt, D. (2017). What Explains Personality Covariation? A Test of the Socioecological Complexity Hypothesis. *Social Psychological and Personality Science,* 8: 943-952.

Lynn, R. (2015). *Race Differences in Intelligence: An Evolutionary Analysis.* Augusta, GA: Washington Summit Publishers.

Lynn, R. (2011a). *Dysgenics: Genetic Deterioration in Modern Populations. 2nd Edition.* London: Ulster Institute for Social Research.

Lynn, R. (2011b). *The Chosen People: A Study of Jewish Intelligence and Achievement.* Augusta, GA: Washington Summit Publishing.

Lynn, R. & Vanhanen, T. (2012). *Intelligence: A Unifying Construct for the Social Sciences.* London: Ulster Institute for Social Research.

Lynn, R. & Vanhanen, T. (2002). *IQ and the Wealth of Nations.* Westport, CT: Praeger.

MacArthur, B. (2006). *Surviving the Sword: Prisoners of the Japanese, 1942-45.* London: Abacus Books.

Malm, S. (14th January 2019). 'It is unacceptable that people granted asylum here have brought evil': Finnish President expresses his 'disgust' at migrant grooming gangs as country's child sex scandal escalates. *Mail Online*, https://www.dailymail.co.uk/news/article-6589765/Finnish-President-expresses-disgust-migrant-grooming-gangs-child-sex-scandal-escalates.html

Mannan, A. (2002). *Stratigraphic evolution and geochemistry of the Neogene Surma Group, Surma Basin, Sylhet, Bangladesh*. PhD Thesis: University of Oulu.

Marusic, A. (2005). History and geography of suicide: Could genetic risk factors account for the variation in suicide rates? *American Journal of Medical Genetics,* 15: 43-47.

McAdams, D. & Pals, J. (2006). A new Big Five: Fundamental principles for an integrative science of personality. *American Psychologist,* 61: 204-217.

Meinilä, M., Finnilä, S. & Majamaa, K. (2001). Evidence for mtDNA Admixture between the Finns and the Saami. *Human Heredity,* 52:160-170.

Meisenberg, G. (2015). Do we have valid country-level measures of personality? *Mankind Quarterly,* 55: 360-382.

Metzen, D. (2012). *The Causes of Group Differences in Intelligence Studied Using the Method of Correlated Vectors and Psychometric Meta-Analysis*. Master's Thesis: University of Amsterdam.

Mitsui, H. (2012). Moomin Troll and Finland in Japan's Social Imagination. *Suomen Antropologi,* 37: 5-21.

MTV. (20th February 2019). Oulun seksuaalirikosvyyhti paisuu yhä, tutkinnassa jo 29 alaikäisiin kohdistunutta juttua – laajimman raiskausjutun lapsiuhria käytettiin hyväksi paljon luultua pitempään. https://www.mtvuutiset.fi/artikkeli/oulun-seksuaalirikosvyyhti-paisuu-yha-tutkinnassa-jo-29-alaikaisiin-kohdistunutta-juttua-laajimman-raiskausjutun-lapsiuhria-kaytettiin-hyvaksi-paljon-luultua-pitempaan/7291024

MTV. (10th February 2018). Uutissuomalainen: Sisäministerinä aloittava Mykkänen: Suomi voisi ottaa jopa 10 000 kiintiöpakolaista. https://www.mtvuutiset.fi/artikkeli/lannen-media-sisaministerina-aloittava-mykkanen-suomi-voisi-ottaa-jopa-10-000-kiintiopakolaista/6764612

Muir, S. (2013). Modes of Displacement: Ignoring, Understating, and Denying Antisemitism in Finnish Historiography. In Muir, S. & Worthen, H. (Eds.). *Finland's Holocaust: Silence of History*. London: Palgrave Macmillan.

Mukwege, D., Mohamed-Ahmad, O. & Fitchett, J. (2010). Rape as a Strategy of War in the Democratic Republic of the Congo. *International Health,* 2: 163-164.

Müller, O. (2016). *The Social Significance of Religion in the Enlarged Europe: Secularization, Individualization and Pluralization.* London: Routledge.

MV??!! Media. (12th May 2017). Oulussa asunut Isiksen jihadisti Taz Rahman kuollut Irakissa – miehen appi on oululainen imaami. https://mvlehti.net/2017/05/12/oulussa-asunut-isiksen-jihadisti-taz-rahman-kuollut-irakissa-miehen-appi-on-oululainen-imaami/

Nahouza, N. (2018). *Wahhabism and the Rise of the New Salafists: Theology, Power and Sunni Islam.* London: I.B. Tauris.

Nenye, V., Munter, P., Wirtinen, T. & Birks, C. (2015). *Finland at War: The Winter War 1939–40.* Oxford: Bloomsbury.

Nettle, D. (2009). Social class through the evolutionary lens. *The Psychologist,* 22: 11.

Nettle, D. (2007). *Personality: What Makes You Who You Are.* Oxford: Oxford University Press.

Neuliep, J., Chaudoir, M., McCroskey, J. (2001). A cross-cultural comparison of ethnocentrism among Japanese and United States college students. *Communication Research Reports,* 18: 137-146.

Norenzayan, A., Gervais, W. & Trzesniewski, G. (2012). Mentalizing Deficits Constrain Belief in a Personal God. *PLOS ONE.* http://journals.plos.org/plosone/article?id=10.1371/journal.pone.0036880

Norenzayan, A. & Shariff, A. (2008). The origin and evolution of religious pro-sociality. *Science,* 322: 58-62.

Norio, R. (2003). Finnish Disease Heritage II: population prehistory and genetic roots of Finns. *Human Genetics,* 112: 457-469.

Oikeusministeriö. (3rd May 2017). Oulu. https://tulospalvelu.vaalit.fi/KV-2017/fi/val564.html

Patil, V. (2015). *Social Problems in India.* Solapur: Laxmi Books.

Pentillä, R. (1992). Official Religions. *Books From Finland,* 1.

Piffer, D. (2019). Evidence for recent polygenic selection on educational attainment and intelligence inferred from GWAS hits: a replication of previous findings using recent data. *PsyArXiv* https://psyarxiv.com/n2yfw/

Piffer, D. (2018). Correlation between PGS and environmental variables. *RPubs,* https://rpubs.com/Daxide/377423

Piffer, D. (2016). Evidence for Recent Polygenic Selection on Educational Attainment Inferred from GWAS Hits. *Preprints,* doi:10.20944/preprints201611.0047.v1

Pinker, S. (18th June 2012). The false allure of group selection. *The Edge,* https://www.edge.org/conversation/the-false-allure-of-group-selection

Post, F. (1994). Creativity and psychopathology. *British Journal of Psychiatry,* 165: 22-34.

Press Association. (17th January 2006). BNP leader 'warned of multiracial hell hole.' *Guardian,* https://www.theguardian.com/uk/2006/jan/17/thefarright.politics

Putnam, R. (2007). *E Pluribus Unum*: Diversity and community in the twenty-first century. The 2006 Johan Skytte Prize lecture. *Scandinavian Political Studies,* 30: 137–174.

Raitio, M., Lindroos, K., Laukkanen, M., Pastinen, T., Sistonen, P., Sajantila, A. & Syvänen, A.-C. (2001). Y-Chromosomal SNPs in Finno–Ugric-Speaking Populations Analyzed by Minisequencing on Microarrays. *Genome Array,* 11: 471-482.

Raunio, A. (2018). Lutheran Impact in the Nordic Socio-Ethical Culture. In Assel, H., Steiger, J. & Walter, A. (Eds.). *Reformatio Baltica: Kulturwirkungen der Reformation in den Metropolen des Ostseeraums.* Berlin: Walter de Gruyter.

Raunio, T. & Tiilikainen, T. (2004). *Finland in the European Union.* London: Routledge.

Read, J. (2010). Can poverty drive you mad? Schizophrenia, socio-economic status and the case for primary prevention. *New Zealand Journal of Psychology,* 39: 7-19.

Relander, J. (2001). How to Kow-Tow. *Books From Finland,* 3.

Reuter, O. (1889). *La Finlande et les Finlandais.* Helsinki: Society of Tourists in Finland.

Rindermann, H. (2007). The g-factor of international cognitive ability comparisons: The homogeneity of results in PISA, TIMSS, PIRLS, and IQ tests across nations. *European Journal of Personality, 21*, 667-706.

Rushton, J.P. (2005). Ethnic nationalism, evolutionary psychology and Genetic Similarity Theory. *Nations and Nationalism,* 11: 485-507.

Rushton, J.P. (1995). *Race, Evolution and Behavior: A Life History Perspective.* New Brunswick, NJ: Transaction Publishing.

Saarelainen, T. (2016). *The White Sniper: Simo Häyhä.* Oxford: Casemate.

Salmela, E. (2012). *Genetic Structure in Finland and Sweden: Aspects of Population History and Gene Mapping.* Doctoral Thesis: Helsinki University.

Salmela, E., Lappalainen, T., Fransson, I., Andersen, P.M., Dahlman-Wright, K., et al. (2008) Genome-Wide Analysis of Single Nucleotide Polymorphisms Uncovers Population Structure in Northern Europe. *PLoS ONE* 3(10): e3519. doi:10.1371/journal.pone.0003519

Salter, F. (2007). *On Genetic Interests: Family, Ethnicity and Humanity in an Age of Mass Migration.* New Brunswick: Transaction Publishers.

Salter, F. & Harpending, H. (2015). J. P. Rushton's Theory of Ethnic Nepotism. In Nyborg, H. (Ed.). *The Life History Approach to Human Differences.* London: Ulster Institute for Social Research.

Service, R. (1995). *Lenin: A Political Life: Volume 2: Worlds in Collision.* London: MacMillan.

Schmitt, D., Allik, J., MacIntyre, R. & Bennet-Martinez, V. (2007). The geographic distribution of the Big Five personality traits: Patterns of profiles of human self-description across 56 Nations. *Journal of Cross-Cultural Psychology*, 38: 173-212.

Scruton, R. (2002) *The Meaning of Conservatism.* South Bend, IN: St. Augustine's Press.

Sela, Y., Shackelford, T. & Liddle, J. (2015). When religion makes it worse: Religiously motivated violence as a sexual selection weapon. In Sloane, D. & Van Slyke, J. (Eds). *The Attraction of Religion: A New Evolutionary Psychology of Religion.* London: Bloomsbury.

Shaw, A. (1915). Finland: The Russian Program and the Working of Woman Suffrage. In *American Review of Reviews*. Vol. 51.

Shehabuddin, E. (2008). *Reshaping the Holy: Democracy, Development, and Muslim Women in Bangladesh*. New York: Columbia University Press.

Silvennoinen, O. (2011). Limits of Intentionality: Soviet Prisoners-of-War and Civilian Internees in Finnish Custody. In Kinnunen, T. & Kivimäki, V. (Eds.). *Finland in World War II: History, Memory, Interpretations*. Leiden: BRILL.

Simonton, D.K. (2003). Exceptional creativity across the life span: the emergence and manifestation of creative genius. In Shavinina, L.V. (Ed.), *The International Handbook of Innovation*. New York NY: Pergamon Press.

Sipilä, J. (2018). *The Dark Side of Helsinki*. Helsinki: Tammi.

Smith, A. & Scheidner, B. (2000). The inter-ethnic friendships of adolescent students: A Canadian study. *International Journal of Intercultural Relations*, 24: 247-258.

Sng, O., Neuberg, S., Varnum, N. & Kenrick, D. (2017). The Crowded Life is a Slow Life: Population Density and Life History Strategy. *Journal of Personality and Social Psychology*, 112: 736-754.

Staff and Agencies. (10th November 2006). BNP leader cleared of race hate charges. *Guardian*, https://www.theguardian.com/politics/2006/nov/10/thefarright.uk

Sunstein, C. (2018). *#Republic: Divided Democracy in the Age of Social Media*. Princeton, NJ: Princeton University Press.

Suomen Uutiset. (2nd March 2019). Suomessa vieraileva brittiläinen islamkriitikko Tommy Robinson suljettiin ulos somesta – tähtää seuraavaksi UKIP-puolueen johtoon? https://www.suomenuutiset.fi/suomessa-vieraileva-brittilainen-islamkriitikko-tommy-robinson-suljettiin-ulos-somesta-tahtaa-seuraavaksi-ukip-puolueen-johtoon/

Suomen Uutiset. (11th January 2019). Oulun tapahtumat kauhistuttavat suomalaisia. https://www.suomenuutiset.fi/oulun-tapahtumat-kauhistuttavat-suomalaisia/

Suvisaari, J., Opler, M., Lindbohm, M.-L. & Sallmen, M. (2014). Risk of schizophrenia and minority status: A comparison of the

Swedish-speaking minority and the Finnish-speaking majority in Finland. *Schizophrenia Research,* 159: 303-308.

Taagepera, R. (2000). *The Finno-Ugric Republics and the Russian State.* London: Routledge.

Taavitsainen, J.-P. (2017). Finland. In Crabtree, P. (2017). *Medieval Archaeology: An Encyclopaedia.* London: Routledge.

Tabor, A., Milfont, T. & Ward, C. (2015). The Migrant Personality Revisited: Individual Differences and International Mobility Intentions. *New Zealand Journal of Psychology,* 44: 89-95.

Taira, T. (28th November 2015). Finland: recent trends and patterns in religion, secularism and atheism. *Observatoire de Religions et de la Laïcité,* http://www.o-re-la.org/index.php/analyses/item/1424-finland-recent-trends-and-patterns-in-religion-secularism-and-atheism

Talve, I. (1997). *Finnish Folk Culture.* Helsinki: Finnish Literature Society.

The Jolly Heretic. (11th February 2019). The Rise of the Mutants: Social Epistasis Amplification, with Michael Woodley of Menie. https://www.youtube.com/watch?v=jXxifvpVHT4

The Times. (5th January 2011). Revealed: conspiracy of silence on UK sex gangs.

Thornhill, R. & Palmer, C. (2001). *A Natural History of Rape: Biological Bases of Sexual Coercion.* Cambridge, MA: The MIT Press.

Trotter, W. (2013). *Frozen Hell: The Russo-Finnish Winter War of 1939-40.* New York: Algonquin Books.

Tulviste, T., Mizera, L., de Geer, B. & Trygvassan, M.-T. (2003). A silent Finn, a silent Finno-Ugric or a silent Nordic? A comparative study of Estonian, Finnish and Swedish Mother Interaction Techniques. *Applied Psycholinguistics,* 24: 249-265.

Turun Sanomat. (1st November 2005). Julma raiskaus toi vankeutta kolmelle Oulussa. https://www.ts.fi/uutiset/kotimaa/1074079095/Julma+raiskaus+toi+vankeutta+kolmelle+Oulussa

Tweedie, Mrs A. (1898). *Through Finland in Carts.* London: Adam and Charles Black.

Ukkonen, R. (16th September 2017**).** Homoparin vihkinyt ja tuomiokapitulin tutkintaan joutunut pappi: Miten kirkko voi

toimia omia sääntöjään ja Suomen lakia vastaan? *YLE,*
https://yle.fi/uutiset/3-9832773

Ulsuner, E. (2000). Trust, Democracy and Governance. European Consortium for Political Research (ECPR) Workshop. Copenhagen.

Uusitalo, H. (5th December 2018). Oulun johtavat poliitikot huolissaan seksuaalirikoksista – Valtuuston puheenjohtaja Hänninen syyllistää suoraan maahanmuuttopolitiikan, Pitko ja Parkkinen eri linjoilla. *Kaleva,* https://www.kaleva.fi/uutiset/oulu/oulun-johtavat-poliitikot-huolissaan-seksuaalirikoksista-valtuuston-puheenjohtaja-hanninen-syyllistaa-suoraan-maahanmuuttopolitiikan-pitko-ja-parkkinen-eri-linjoilla/811542/

Van der Linden, D., Dutton, E. & Madison, G. (2018). National-Level indicators of androgens are related to the global distribution of number of scientific publications and science Nobel prizes. *Journal of Creative Behavior,* doi.org/10.1002/jocb.351

Van der Linden, D., te Nijenhuis, J. & Bakker, A. (2010). The General Factor of Personality: A meta-analysis of Big Five inter-correlations and a criterion related validity study. *Journal of Research in Personality*, 44: 315–327.

Vanhanen, T. (2012). *Ethnic Conflicts: Their Biological Roots in Ethnic Nepotism.* London: Ulster Institute for Social Research.

Virtanen, A.-M. (1985). Gobineau's Racial Doctrine and the Idealization of Germanic Peoples. In Kemiläinen, A. (Ed.). *Mongols or Germanics?* Helsinki: Finnish Historical Society.

Voice of Europe. (11th December 2018). Media cover up: Middle Eastern men groom and rape girls as young as 10 in Finland. http://voiceofeurope.com/2018/12/media-cover-up-middle-eastern-men-groom-and-rape-girls-as-young-as-10-in-finland/

Voice of Europe. (14th May 2018). 93% of migrant sex crimes in Finland are committed by migrants from Islamic countries – Study. https://voiceofeurope.com/2018/05/93-of-migrant-sex-crimes-in-finland-are-committed-by-migrants-from-islamic-countries-study/

Warren, M. (2018). Trust and Democracy. In Ulsuner, E.M. (Ed.). *The Oxford Handbook of Social and Political Trust.* Oxford: Oxford University Press.

Weinberg, R., Scarr, S. & Waldman, I. (1992). The Minnesota trans-racial adoption study: A follow-up of IQ test performance at adolescence. *Intelligence,* 16: 117-135.

Wiik, K. (2008). Where did European Men Come From? *Journal of Genetic Genealogy,* 4: 35-85. http://www.jogg.info/41/Wiik.pdf

Wiik, K. (2006). *Mista suomalaiset ovat tulleet?* Tampere: Pilot Publications.

Will, G.F. (1994). *The Leveling Wind: Politics, the Culture, and Other News, 1990-1994.* New York: Viking.

Wilson, W. (1976). *Folklore and Nationalism in Modern Finland.* Bloomington: Indiana University Press.

Withnall, A. (6th September 2015). Refugee crisis: Finland's Prime Minister pledges to give up his home to accommodate refugees. *Independent.*

Woodley, M.A. (2011a). The cognitive differentiation-integration effort hypothesis: A synthesis between the fitness indicator and life history models of human intelligence. *Review of General Psychology,* 15: 228-245.

Woodley, M.A. (2011b). Heterosis doesn't cause the Flynn effect: A critical examination of Mingroni. *Psychological Review,* 118: 689-693.

Woodley of Menie, M.A., Saraff, M., Pestow, R. & Fernandes, H. (2017a). Social Epistasis Amplifies the Fitness Costs of Deleterious Mutations, Engendering Rapid Fitness Decline Among Modernized Populations. *Evolutionary Psychological Science,* 3: 181-191.

Woodley of Menie, M.A., Cabeza de Baca, T., Fernandes, H. et al. (2017b). Slow and Steady Wins the Race: K Positively Predicts Fertility in the USA and Sweden. Evolutionary Psychological Science, 3: 109-117.

Wuorinen, J. (2015). *Finland and World War II, 1939-1944.* Pickle Partners.

Virtaranta-Knowles, K., Sistonen, P. & Nevanlinna, H. (1991). A population genetic study in Finland: Comparison of the Finnish- and Swedish-speaking populations. *Human Heredity,* 41: 248-264.

YLE. (13th February 2019). Yet another suspected sexual abuse case in Oulu.

https://yle.fi/uutiset/osasto/news/yet_another_suspected_sexual_abuse_case_in_oulu/10644762

YLE. (12th February 2019). Two new suspected cases of child sexual abuse in Oulu. https://yle.fi/uutiset/osasto/news/two_new_suspected_cases_of_child_sexual_abuse_in_oulu/10642570

YLE. (15th June 2018). Finland's first terror attack: Life sentence for Turku stabber. https://yle.fi/uutiset/osasto/news/finlands_first_terror_attack_li fe_sentence_for_turku_stabber/10257371

YLE. (5th September 2015). Dozens of asylum seeker protesters cross Finnish border to Sweden. https://yle.fi/uutiset/osasto/news/dozens_of_asylum_seeker_pr otesters_cross_finnish_border_to_sweden/9816823

YLE. (24th November 2015). Murrosikäinen tyttö raiskattiin Kempeleessä - poliisi ottanut kiinni kaksi ulkomaalaista nuorta miestä. https://yle.fi/uutiset/3-8477391

YLE. (11th July 2013). Räsänen's comments cause spike in church resignations. https://yle.fi/uutiset/osasto/news/rasanens_comments_cause_sp ike_in_church_resignations/6728183

Zerjal, T., Dashnyam, B., Pandya, A., Kayser, M., Roewer, L. et al. (1997). Genetic relationships of Asians and Northern Europeans, revealed by Y-chromosomal DNA analysis. *American Journal of Human Genetics,* 60: 1174-1183.

Made in the USA
San Bernardino, CA
04 September 2019